AQA Design and Technology

Food Technology

GCSE

Jane Girt

Jenny Hotson

Garry Littlewood

ornes

Published in 2009 by:
Nelson Thornes Ltd
Delta Place
27 Bath Road
CHELTENHAM
GL53 7TH
United Kingdom

09 10 11 12 13 / 10 9 8 7 6 5 4 3 2 1

A catalogue record for this book is available from the British Library

ISBN 978 1 4085 0272 3

Cover photograph by Nelson Thornes
Illustrations by Roger Penwill, Angela Knowles, Harry Venning and Fakenham Photosetting
Page make-up by Fakenham Photosetting

Printed and bound in Spain by GraphyCems

Acknowledgements:

Nelson Thornes would like to thank the following companies and individuals who assisted with, or supplied photographs which appear in this book: Alamy; Apex Bakery Equipment Company, Inc; Assured Food Standards (AFS); Benjamin Phillips Photography; China National Light Industry Council; Deliciously Dessert; DK Images European Commission; Food and Drink photos; fusetek; Georgia Tech Photo: Gary Meek; Greencore Soups and Sauces and Greencore Cakes and Desserts (part of Greencore plc); Indexing Specialists (UK) Ltd; Health and Safety Executive; iStockphoto; J Sainsbury PLC; Nathan Allan Photography; Pippa Drake-Lee; Science Photo Library; Shutterstock; The Institute of Mathematical Sciences (Chennai, India); sz-olong (China); Tesco PLC.

Special appreciation is offered to Clive Pryce; Jessica King; Katherine Wheeler; Sophie Brooks; and the pupils and teachers of Banbury School, Oxfordshire. Photo Research and commissioned photography by Ann Asquith; Dora Swick; Jason Newman; Shane Lapper; Tara Roberts; of Image Asset Management & unique dimension.com.

Contents

4

Nelson Thornes has worked in partnership with AQA to ensure this book and the accompanying online resources offer you the best support for your GCSE course.

All resources have been approved by senior AQA examiners so you can feel assured that they closely match the specification for this subject and provide you with everything you need to prepare successfully for your exams.

These print and online resources together **unlock blended learning**; this means that the links between the activities in the book and the activities online blend together to maximise your understanding of a topic and help you achieve your potential.

These online resources are available on kerboodle! which can be accessed via the internet at **http://www.kerboodle.com/live**, anytime, anywhere. If your school or college subscribes to kerboodle! you will be provided with your own personal login details. Once logged in, access your course and locate the required activity.

For more information and help on how to use kerboodle! visit **http://www.kerboodle.com**.

How to use this book

Objectives

Look for the list of **Learning Objectives** based on the requirements of this course so you can ensure you are covering everything you need to know for the exam.

AQA Examiner's tip

Don't forget to read the **AQA Examiner's Tips** throughout the book as well as practice answering **Examination-style Questions**.

Visit **http://www.nelsonthornes.com/aqagcse** for more information.

AQA examination-style questions are reproduced by permission of the Assessment and Qualifications Alliance.

What is Food Technology?

Food Technology is a very exciting, creative and interesting subject to study. There is much to learn but most of this will involve 'learning-by-doing'. You will learn a tremendous amount through making activities which involve working with food ingredients and food products. Hopefully this is one of the reasons you have chosen to study GCSE Food Technology. Food Technology involves:

- learning about foods, ingredients, processes, techniques
- experimenting, investigating and testing products
- being creative and designing new products
- understanding how a product is developed in a test kitchen
- having an insight into how products are made in industry
- developing skills which enable you to make food products
- understanding about diets and health
- making choices as consumers.

Food Technology involves you working in the same way as a food technologist works in industry. The food technology room becomes the test kitchen. You will be designing and making a wide range of products, developing your making skills but also have the exciting opportunity to design and make different food products. We call this the design process. This is explained in more detail at the beginning of the Designing and making practice unit.

Your GCSE grade will be awarded as a result of completing two units of work:

Unit 1

A written examination worth 40% of the total marks which will require you to apply what you have learned during the course in an examination situation.

Unit 2

A coursework project called Design and making practice which involves answering a design task and designing and making a food product; this is worth 60% of the final mark.

In both your designing and making practice and the written examination, you will be assessed on how you demonstrate your knowledge skills and understanding.

Food Technology will also help you to develop many other important and valuable skills. These include:

- organisation skills
- life skills
- independent skills as well as team working
- information and communication skills.

What is this book about?

You will want to achieve a good GCSE Food Technology grade. This book has been written to support your learning. You will see that the book provides you with clear concise explanations, descriptions and examples to support you through the course and also helps prepare you for your Design and making practice and the written examination. It has been written as a student book to provide you with hints and tips as you go through the GCSE course.

The book is divided into two units which match the AQA GCSE Food Technology Specification:

Unit 1: Written paper

- Materials and components
- Processes and manufacture
- Design and market influences

Unit 2: Design and making practice

In addition to information the book will also provide examples of examination-style questions at the end of written paper sections.

Unit 1 called the 'Written paper' includes all the knowledge and understanding you need for the written examination, but you also require this knowledge and understanding for Unit 2 'Designing and making practice'.

Food Technology and a future career

Studying Food Technology can lead to exciting and well-paid career opportunities in the food industry. The food industry is expanding all the time and is one of the largest employers in the UK. Food technologists are much sought after and it is estimated there are three jobs for every graduate leaving University. Having a Food Technology qualification can lead to careers in: food marketing, product development, diet–related industries, and more.

In conclusion

As you focus on different topics with your teacher this book will help you gain the knowledge, understanding and skills you need to be successful, so learn and enjoy.

The controlled assessment tasks in this book are designed to help you prepare for the tasks your teacher will give you. The tasks in this book are not designed to test you formally and you cannot use them as your own controlled assessment tasks for AQA. Your teacher will not be able to give you as much help with your tasks for AQA as we have given with the tasks in this book.

Materials and components

Design and Technology involves the study of materials and components. The materials and components we use in Food Technology are ingredients. When studying food we need to understand the functional properties of the ingredients that we use. The majority of this can be learnt by carrying out making activities/investigations and experiments in a test kitchen which is both interesting and exciting.

To be able to produce high quality and tasty food products you need to have a good understanding and working knowledge of the ingredients you are using. You will be given the opportunity to gain knowledge and understanding of the functions, working characteristics and processing techniques when designing and making food products.

■ What will you study in this section?

After completing Materials and components (Chapters 1–5) you should have a good understanding of:

- the functional properties of ingredients
- the nutritional properties of food
- how to combine ingredients. This will help you to use ingredients successfully
- the effects of acids and alkalis
- the standard components, ready made ingredients, used in food processing.

■ Making/cooking

One of the reasons you will have chosen GCSE Food Technology is to cook. The best way to learn about ingredients is to be engaged in making activities. Throughout each chapter activities have been suggested to allow you to learn the theoretical aspects of the unit by 'learning-by-doing'. It is important that you can explain why:

- a sauce thickens
- a chilled dessert sets
- bread rises
- pastry has a crumbly texture
- oil and water do not mix
- low fat spreads are not suitable to make a cake.

You must also be able to explain how to increase the amount of fibre (NSP) in food products.

How will you use the information?

When you have studied this section it is very important to learn the work thoroughly. You will be tested on some of this knowledge in the examination. It is also essential that you demonstrate your knowledge of ingredients throughout the controlled assessment. Whenever you are engaged in making activities try to explain the functions of ingredients you are using.

Working with different ingredients and understanding how ingredients and methods work together will result in the production of successful and quality food products in addition to developing excellent making/cooking skills.

1 The functions and properties of food

1.1 Starch

What are carbohydrates?

Two types of carbohydrates are:

- starches – which are found in flour, potatoes, pasta, rice and bread
- sugars – which are found in fruits, drinks and sweet baked products.

Types and uses of starch

Starch is obtained from cereals such as wheat and maize.

- Wheat flour contains a high percentage of starch.
- Cornflour and arrowroot are pure starch. Cornflour is made from maize and arrowroot is made from a tropical root.

Starch has many useful functions and properties, for example:

- bulking agent
- thickening agent
- gelling agent.

Bulking agent

Flour is often the main ingredient in many food products and we call this the bulk ingredient. The starch forms the main structure of the product. Examples include: biscuits, cakes and pastry.

Thickening agent

Raw starch tastes floury and therefore needs to be cooked, for example plain flour. When starch is heated in a liquid, the walls of the starch become soft and allow liquid to pass slowly through them. This makes the granules swell until they burst. This process is called **gelatinisation** and is explained in more detail below.

A *Gelatinisation of starch*

Gelatinisation

- Starch particles do not dissolve in liquid. Instead they form something called a **suspension.**
- If the liquid is not stirred, the starch granules sink to the bottom, stick together and start to form lumps.

B *Examples of starch – cornflour and wheat*

- When heated, at 60°C, the starch granules begin to absorb the liquid and swell.
- At 80°C the particles will have absorbed about five times their volume of water until they burst open and release starch, thickening the liquid. The mixture becomes thick and viscous. This process is gelatinisation.
- Gelatinisation is complete when the liquid reaches boiling point, 100°C.
- When the sauce cools, it goes even thicker, setting into a **gel**.

Examples of gelatinisation

- Wheat flour is used with fat and milk to make a white sauce, e.g. cheese sauce for lasagne.
- Arrowroot is blended with fruit juice to make a glaze for a fruit flan.
- Cornflour is used for sweet and savoury sauces, e.g sweet and sour sauce.

Gelling agent

After full gelatinisation has occurred the thickened liquid forms a gel. On cooling the gel then solidifies and is no longer in a liquid state. When reheated the sauce reverts back to its liquid state. Examples include blancmanges and custards where cornflour is used.

Modified starches

This is a starch that has been altered (modified) to perform additional functions to react to different processes. In food technology these are sometimes called smart starches.

- **Modified starch** is used to thicken food when boiled water is added, for example cup-a-soup.
- Pre-gelatinised starch is used to thicken instant desserts without heat. A cold liquid can be added and the dessert will thicken when stirred or whisked, e.g. instant whip.

Key terms

Gelatinisation: when heated, starch granules absorb liquid and swell and then burst to thicken a liquid.

Suspension: a solid held in a liquid.

Gel: a small amount of a solid mixed in a large amount of liquid that then sets.

Modified starches: starches that have been altered to perform additional functions.

C *Products containing modified starches*

Summary

You should now be able to:

explain the main uses of starch

explain 'gelatinisation'

understand (through practical activities) how different starches can be used to make food products.

Activities

Investigate how different starches thicken a liquid by making different sauces using: plain flour, cornflour and arrowroot.

1 Compare the appearance of the raw and cooked starch through a microscope.

2 Carry out a sensory analysis of the sauces.

3 Write an evaluation of the results.

4 To gain practical experience, make the following:
   a Macaroni cheese or lasagne using plain flour to make a white sauce.
   b Sweet and sour chicken/vegetables using cornflour to make a blended sauce.
   c Fruit flan using arrowroot to make a glaze.

kerboodle!

1.2 Sugars

Where does sugar come from?

Sugar comes from sugar beet or sugar cane.

- Sugar cane grows in tropical countries, e.g. the West Indies. The sugar is stored in the long stems (canes) as juice.
- Sugar beet grows in cooler countries, e.g. the United Kingdom. The sugar is stored in the root.

Types of sugar

There are many different types of sugar used in food production. Sugar is produced by processing the cane or beet. It is crushed and mixed with water and the liquid is boiled to form sugar crystals. Sugar is identified by the size and colour of the crystals.

A *Table to show types of sugar*

Sugar type/ Sugar product	Description	Uses
Granulated	General purpose sugar. White medium sized crystals, coarser than caster sugar	Sweetening drinks, adding to breakfast cereals
Caster	Small white crystals that dissolve quickly	Cakes, biscuits, muffins, meringues
Icing	Fine, white powder that dissolves instantly	Cake decorating and icings: confectionery
Demerara	Large, light brown crystals	Toppings on puddings, hot drinks
Muscovado	Dark brown, fine, sticky crystals with a strong and distinctive flavour	Fruit cakes, puddings, gingerbread

Adapting the sugar in recipes

In response to dietary recommendations and as we have become more aware of healthy eating, the demand for low calorie products has risen. Manufacturers have responded to this and developed sugar substitutes. Manufacturers use sugar substitutes, sometimes called artificial sweeteners:

- to create a 'low calorie' or 'diet' product, e.g. low calorie drinks
- to promote food for dental health, e.g. sugar-free chewing gum
- to market products for people with diabetes. People with diabetes need to control their sugar intake. Examples of diabetic products available are jam and some chocolate products.

Functions of sugar

Think what food would taste like if we did not have sugar. Sugar is not just a sweetener; it can be used in different ways. Many savoury

B *Sugar cane*

∞links

Visit the sugar bureau www.sugar-bureau.co.uk to see the processing of sugar beet and cane.

C *Types of sugar*

The functions of sugar

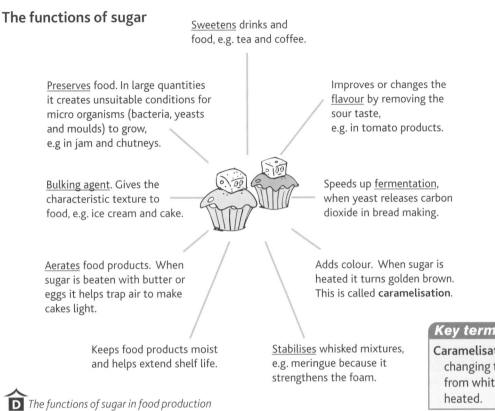

Sweetens drinks and food, e.g. tea and coffee.

Preserves food. In large quantities it creates unsuitable conditions for micro organisms (bacteria, yeasts and moulds) to grow, e.g in jam and chutneys.

Improves or changes the flavour by removing the sour taste, e.g. in tomato products.

Bulking agent. Gives the characteristic texture to food, e.g. ice cream and cake.

Speeds up fermentation, when yeast releases carbon dioxide in bread making.

Aerates food products. When sugar is beaten with butter or eggs it helps trap air to make cakes light.

Adds colour. When sugar is heated it turns golden brown. This is called **caramelisation**.

Keeps food products moist and helps extend shelf life.

Stabilises whisked mixtures, e.g. meringue because it strengthens the foam.

D *The functions of sugar in food production*

> **Key terms**
>
> **Caramelisation:** the process of changing the colour of sugar from white to brown when heated.

products contain sugar. Look at the food label for baked beans and tomato sauce; surprisingly these contain sugar!

Any ingredient ending in 'ose' on an ingredients list is a type of sugar, e.g. fructose is the sugar in fruit, and lactose is the sugar in milk.

Modifying recipes

When we use sugar substitutes it can be difficult to get good results. Sugar substitutes:

- can leave a bitter after-taste
- lack the bulk which is necessary for a recipe
- do not have the same properties of sugar and therefore the results can be unsatisfactory
- reduce the shelf life.

> **Activities**
>
> 1 Investigate the effects of using artificial sweeteners to replace sugar in a cake, biscuit or scone recipe. In small groups make different samples of cakes with different ratios of a sugar substitute.
>
> 2 In groups investigate different types of sugar. Make small samples of scones, cakes or biscuits using: caster, granulated, demerara and muscovado sugar. Produce annotated sketches or digital images to support your findings. What are your conclusions from the investigation?
>
> 3 Make mini carrot cakes to show how carrots can add sweetness to a baked product.

> **Summary**
>
> You should now be able to:
>
> name different sugars and explain their uses
>
> explain the functions of sugar in cooking
>
> explain the effects of sugar substitutes in recipes.

> AQA *Examiner's tip*
>
> Learn the main functions of sugar in different food products and be able to apply this knowledge when making different food products.

1.3 Protein

Proteins

In Food Technology you will use many different protein foods such as milk, yogurt, cheese, meat, fish, eggs, alternative proteins (TVP: soya, tofu, mycro protein), nuts and pulses. All these ingredients perform specific functions when food is cooked.

The science

Protein foods are made up of small units called **amino acids** that are linked together like the links in a chain. When protein foods are heated the links in the chain change and this alters the structure of the food. Think about a raw egg and the changes that take place when an egg is fried or scrambled.

Function and properties of protein foods

A *The functional properties of three protein foods*

Cheese	Milk	Yogurt
Is used to add flavour, texture and colour to sauces and savoury fillings, e.g. quiche Lorraine	Is used to bind other ingredients together	Thickening agent, e.g. salad dressings and dips
When heated melts and is often mixed with other ingredients to provide different textures, e.g. pizza	Is used to give flavour, colour and consistency to products, e.g. Yorkshire pudding, muffins	Adds a creamy texture to soups and sauces
		Adds an acidic flavour

The use of eggs

Eggs are a very versatile ingredient (they have many uses), and are used widely in cooking.

Aeration

The whole egg and egg white on its own are capable of trapping air – this is called **aeration**.

When eggs are whisked, the protein stretches and incorporates air bubbles into the mixture to create a foam, e.g. mousse, Swiss roll. When egg whites are whisked on their own they can hold up to seven times their volume, e.g. meringue.

The heat created by the whisking keeps the foam stable and partially sets the foam. If left to stand, the foam will gradually collapse, but when heated the foam will become solid and permanent.

Objectives

Understand the functions of protein foods in food production.

Understand the main use of eggs in food preparation.

Key terms

Amino acids: the small units that form the chains in protein.

Aeration: when air is trapped in a mixture.

Coagulation: when eggs are heated they change colour and become firm-set.

Emulsion: a mixture of two liquids.

B *Egg whites can hold up to seven times their own volume*

Remember

Raw eggs may contain food poisoning bacteria called Salmonella – eggs must be cooked thoroughly to destroy the bacteria. See Chapter 7.

∞ links

Further explanation of raising agents can be seen on page 28.

Uses of eggs

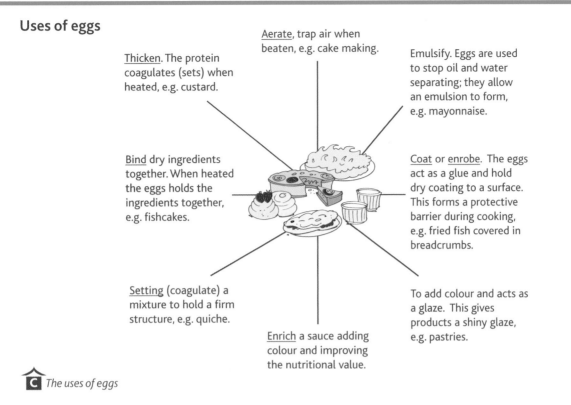

Aerate, trap air when beaten, e.g. cake making.

Thicken. The protein coagulates (sets) when heated, e.g. custard.

Emulsify. Eggs are used to stop oil and water separating; they allow an emulsion to form, e.g. mayonnaise.

Bind dry ingredients together. When heated the eggs holds the ingredients together, e.g. fishcakes.

Coat or enrobe. The eggs act as a glue and hold dry coating to a surface. This forms a protective barrier during cooking, e.g. fried fish covered in breadcrumbs.

Setting (coagulate) a mixture to hold a firm structure, e.g. quiche.

To add colour and acts as a glaze. This gives products a shiny glaze, e.g. pastries.

Enrich a sauce adding colour and improving the nutritional value.

C *The uses of eggs*

Coagulation

When eggs are heated they change from a liquid to a solid and set. This is called **coagulation**. Think about frying an egg: which part sets first?

- Egg white starts to coagulate at 60°C but egg yolk needs to be heated to a higher temperature of 65°C.
- Full coagulation happens at 70°C.
- Coagulation is used to make egg custards and quiches and to give structure to cakes.

Emulsification

When oil and water are forcibly mixed together they form what is called an **emulsion**. After a few minutes the oil and water will separate unless an emulsifier is added.

Egg yolk contains a substance called Lecithin. When egg yolk is mixed with oil and vinegar it stops the liquid separating and is therefore known as an emulsifier. Egg yolk acts as an emulsifier to make mayonnaise and creamed cake mixture. Lecithin is also found in soya beans and this is used as an emulsifier in food manufacturing.

AQA **Examiner's tip**

Always try to use technical terms such as coagulate, emulsify and aerate. Higher marks are available for the use of technical terms.

Activities

1 Research the ethical issues of different methods of egg production.

2 Make a range of dishes to illustrate the functional properties of eggs:

a Aeration: Swiss roll and fruit flan

b Binding: Burgers and fish cakes

c Coagulation: Quiche.

Summary

You should now be able to:

explain the different functions of eggs

explain: aerate, coagulate and emulsify

make good quality products demonstrating aeration and coagulation.

1.4 Fats and oils

Types of fat

There are many different types of fat. Fats are produced from three different sources:

- animals
- fish
- vegetables.

A general rule is that at room temperature fats are solid or semi solid, e.g. butter. Oils are liquid, e.g. olive oil. There are many different fats and oils available to the consumer.

- Butter is made from churning cream. It provides about 80g fat per 100g. Butter is high in saturated fat. Ghee is prepared by heating and clarifying butter and is used in Indian cookery.
- Lard comes from pigs' fat and provides about 99g of fat per 100g. It is softer than butter and white in colour.
- Suet is made from the fat that surrounds animals' organs and has a similar fat content to lard.
- Soft spreads are blended from vegetables oils. The oil is hardened by adding hydrogen gas. Soft spreads provide approximately 80g fat per 100g.
- Vegetable oils are mainly produced from oil seeds, e.g. sunflower seeds as well as from the flesh of some fruits, e.g. olives.
- Reduced or low fat spreads contain 40–80% fat. They have a high water content so are not suitable for some cooking methods. They are an emulsion of water and oil.

Saturated v. unsaturated fats

Saturated fats are found in: lard, butter, suet and some vegetable oils, e.g. coconut milk and some soft spreads. They should be used sparingly as they contain more cholesterol than other fats.

Unsaturated fats are found mainly in vegetable oils, e.g. cooking oils. Unsaturated fats contain less cholesterol.

Shortening

Fats have a **shortening** effect in pastry and biscuit mixtures. This is how it happens:

1 The fat coats the flour particles and this prevents the absorption of water, giving a waterproof coating.
2 This prevents **gluten** from developing which can give mixtures an elastic and stretchy texture. Gluten is formed when the protein in the flour mixes with water. If the gluten cannot form, the mixture is shortened giving a characteristic, short, melt-in-the-mouth and crumbly texture.

The next time you try a piece of shortbread remember how the crumbly texture was formed.

A *Butter and olive oil, two types of fat*

B *The shortening effect*

∞ **links**

Further explanation of emulsification can be seen on page 30.

The functions of fats and oils

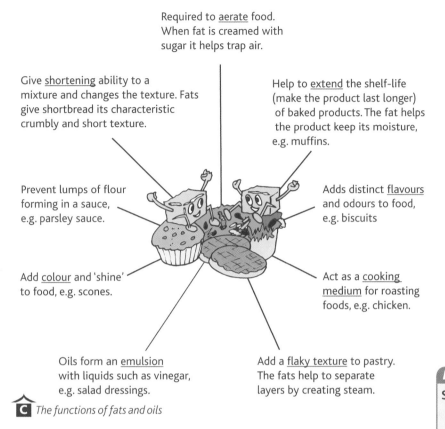

Required to <u>aerate</u> food. When fat is creamed with sugar it helps trap air.

Give <u>shortening</u> ability to a mixture and changes the texture. Fats give shortbread its characteristic crumbly and short texture.

Help to <u>extend</u> the shelf-life (make the product last longer) of baked products. The fat helps the product keep its moisture, e.g. muffins.

Prevent lumps of flour forming in a sauce, e.g. parsley sauce.

Adds distinct <u>flavours</u> and odours to food, e.g. biscuits

Add <u>colour</u> and 'shine' to food, e.g. scones.

Act as a <u>cooking medium</u> for roasting foods, e.g. chicken.

Oils form an <u>emulsion</u> with liquids such as vinegar, e.g. salad dressings.

Add a <u>flaky texture</u> to pastry. The fats help to separate layers by creating steam.

C *The functions of fats and oils*

■ Low fat spread

Too much fat can be a danger to long term health. Low fat spreads can contain about half the fat of soft spreads and butter. They contain higher proportions of water and this is how the fat has been reduced. The high water content makes low fat spreads unsuitable for frying and baking.

Activities

1   Make shortbread or biscuits testing different fats. Use lard, baking fat, butter and low fat spread. Produce a detailed summary of the investigation explaining which fats are suitable for biscuit making, fully justifying your results.

2   Explain how fats have a 'shortening' effect when making biscuits.

Summary

You should now be able to:

name the different types of fats and oils

know the different uses of fats and oils when cooking

explain how fats 'shorten' pastry and biscuits.

Key terms

Shortening: when fat coats the flour particles preventing absorption of water resulting in a crumbly texture.

Gluten: protein found in flour.

Activity

3   Carry out a blind comparison of different fats and oils. Spread a small amount of fat or oil on a cracker or bread and complete a detailed analysis table. Compare the taste, texture, aroma, appearance, spreadability, nutritional value and cost.

AQA *Examiner's tip*

Remember that soft spreads and butter contain similar amounts of fat per 100g. The difference is that butter contains more saturated fat. This type of fat should be reduced to maintain a healthy diet.

kerboodle!

1.5 Processes and techniques

Throughout this chapter there have been many references to the functions of ingredients and why particular ingredients are required to make a successful and good quality food product. Using different processes and cooking techniques is also important.

Objectives

Understand the different processes and techniques used in food preparation and production.

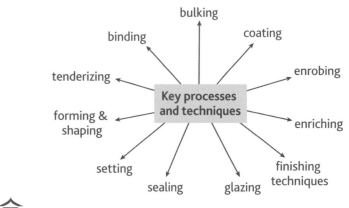

A *Key food technology processes and techniques*

Enrobing

Enrobing is coating one product with another ingredient to give it an outer layer, e.g. breadcrumbs on fish fingers or coating a biscuit with chocolate.

Activity

1   Use a search engine, e.g. Google, to find images of enrobing a food product.

Shaping and forming

Shaping is an important element to creating an attractive food product.

- By hand: this is one of the simplest methods, e.g. meatballs, bread rolls.
- Moulds: shapes can be created by using a mould, e.g. jelly.
- Extruding: forcing a mixture through a nozzle, e.g. cake icings, pasta.
- Equipment: using cutters, presses and tins to add a shape to a food product.
- Rolling: the pliable texture of dough allows different shapes to be created, e.g. cornish pasties.

In industry these techniques are recreated on a large scale to allow thousands of products to be made in a short space of time.

Finishing techniques

The appearance of a food product is essential. To make both sweet and savoury products look attractive you could use the following **finishing** techniques:

- **piping**: fresh cream, chocolate and mashed potato can be piped using bags and nozzles to create an attractive and professional design.

links

See Chapter 9 Food production.

Activity

2   Make a range of food products to demonstrate different finishing techiniques.

- **dusting**: icing sugar can be dusted over products using a dredger or a sieve.

- **glazing**: a glaze is a smooth shiny coating which gives an attractive finish, e.g. jam can be warmed and used to cover a fruit flan.

- **egg-wash glazing**: a mixture of milk and egg brushed onto pastry before cooking gives a shiny golden finish.

- **garnishes**: herbs, such as parsley, and fruit such as fanned strawberries can enhance the finish of dishes.

- **chocolate**: chocolate swirls, grated chocolate and other chocolate shapes can add interest to a dessert.

- **icings**: different icings can be added to sweet baked products such as: butter cream, glace icing, fondant icing, fudge icing, etc. to create a professional finish.

- **pastry dishes**: pastry can be prepared in different ways, e.g. a lattice top.

B *An example of a piping technique*

C *Examples of different finishing techniques*

links

For further examples of different finishing techniques refer to Chapter 16.

Tenderising

Tough meats must be tenderised to make them easier to digest and more enjoyable to eat. The aims of tenderising are to break down or soften the connective tissue or muscle in meat. Methods of tenderising include:

- marinating: in lemon juice, vinegar and wine
- mechanical: using a mallet or chopping
- cooking: slow and moist cooking can also tenderise meat.

Activities

3 Produce and annotate a mood board of different finishing techniques that have enhanced the presentation of dishes.

4 Carry out research to find out why fish does not require tenderising.

Summary

You should now be able to:

explain key food technology terms relating to food processing and techniques

demonstrate some of the techniques when making food products.

AQA *Examiner's tip*

- Learn about the functions of ingredients, processes and techniques; experiment making different products.

- Always aim to present the finished product to a high standard; this will be essential when presenting dishes for Unit 2 Design and making practice.

Key terms

Enrobing: coating and surrounding a product with another ingredient.

Finishing: completing the presentation of a food product to a high standard.

2.1 Diet and health

What is a balanced diet?

Food is vital to health. Having a variety of foods in our diet helps us to enjoy a healthy life. Food provides us with nutrients, which are essential to keep us fit and healthy. A balanced diet provides all the necessary nutrients in the appropriate proportions and quantities to meet the body's needs. To follow a balanced diet we must make sure we eat a variety of foods.

The human body is like a complex piece of machinery in that it is prone to faults and weaknesses if it is not treated correctly. This can happen if too little or too much food is consumed, or if we eat an unbalanced diet.

The eatwell plate

The **eatwell plate** is the healthy eating model for the United Kingdom. It is made up of five different groups and shows the balance and variety of foods we should include in our diet. The eatwell plate makes healthy eating easier to understand by showing the types and proportions of foods we need to have a healthy and well balanced diet. The two keys to a healthy diet are:

- eating the right amount of food for how active you are
- eating a range of foods to make sure you are getting a balanced diet.

Objectives

Understand the need for a balanced diet.

Understand nutritional advice.

Understand how diet affects health.

Remember

People sometimes think that the word 'diet' refers only to trying to lose weight. This is not correct. Diet is a general term describing the food we eat.

Key terms

Eatwell plate: a healthy eating model, to encourage people to eat the correct proportions of food to achieve a balanced diet.

○○ links

To see a larger version of the eatwell plate and further explanation of the food groups visit www.eatwell.gov.uk/healthydiet/eatwellplate.

The eatwell plate

FOOD STANDARDS AGENCY
food.gov.uk

Use the eatwell plate to help you get the balance right. It shows how much of what you eat should come from each food group.

Fruit and vegetables

Bread, rice, potatoes, pasta and other starchy foods

Meat, fish, eggs, beans and other non-dairy sources of protein

Foods and drinks high in fat and/or sugar

Milk and dairy foods

A *The eatwell plate*

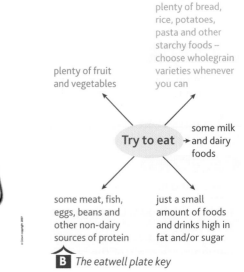

plenty of fruit and vegetables

plenty of bread, rice, potatoes, pasta and other starchy foods – choose wholegrain varieties whenever you can

Try to eat → some milk and dairy foods

some meat, fish, eggs, beans and other non-dairy sources of protein

just a small amount of foods and drinks high in fat and/or sugar

B *The eatwell plate key*

1   Examine graph **C** – Obesity levels in the UK. Carry out research to find out why obesity levels are increasing. Produce a handout or PowerPoint presentation to present your findings.

2   Design and make a product or meal for a child that follows the Guidelines for a healthy diet/Eating '5 a day'.

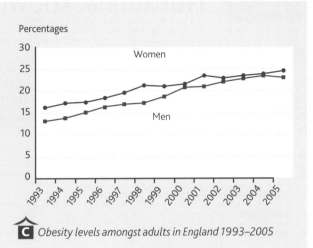

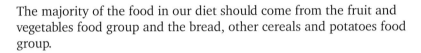

C *Obesity levels amongst adults in England 1993–2005*

The majority of the food in our diet should come from the fruit and vegetables food group and the bread, other cereals and potatoes food group.

Guidelines for a healthy diet

The eatwell plate is based on the Government's '8 guidelines for a Healthy Diet', which are:

1   Base your meals on starchy foods, e.g. potatoes, pasta, rice, bread
2   Eat lots of fruit and vegetables
3   Eat more fish
4   Cut down on saturated fat and sugar
5   Try to eat less salt – no more than 6g a day
6   Get active and try to be a healthy weight
7   Drink plenty of water
8   Don't skip breakfast.

Five a day

An 'unhealthy diet' could be one which contains high quantities of fats, sugars and salt and low amounts of non starch polysaccharide (dietary fibre). Such a diet can lead to:

- obesity
- strokes
- high blood pressure
- coronary heart disease
- cancers
- tooth decay
- diabetes: Type 2.

Scientific studies have shown that people who eat a lot of fruit and vegetables may have a lower risk of ill health. For this reason, it is recommended that you eat at least five portions of fruit and vegetables every day.

Remember

Eat a range of foods from the five food groups to make sure you have a balanced diet. But most of all ENJOY your food.

AQA   Examiner's tip

- Be familiar with the eatwell plate and understand the types and proportions of foods we need to have a healthy and well balanced diet.
- Know the diseases associated with eating an unbalanced diet.

Summary

You should now be able to:

explain why it is important to have a balanced diet

make healthy choices when selecting food to eat

name the diseases associated with an unbalanced diet.

2.2 Nutritional knowledge

What are nutrients?

Nutrients are substances that perform different functions in the body. No single food will provide all the essential nutrients that the body needs to be healthy and function efficiently. There are five main groups of nutrients:

- protein
- fat
- carbohydrates (sugar and starches)
- vitamins
- minerals.

There are two other non-nutrients that are also needed: water and **non starch polysaccharides** (dietary fibre).

Functions of nutrients

Each nutrient has a specific function in the body.

Objectives

Understand the functions and sources of nutrients and non starch polysaccharide.

Understand the nutritional needs of different dietary groups.

Key terms

Nutrient: the part of a food that performs a particular function in the body.

Non starch polysaccharide: the part of food that is not digested by the body.

A *Functions and sources of nutrients*

Nutrient	Job (function) in the body	Sources
Carbohydrates (sugar and starch)	Gives the body energy	Sugar, honey, jam
		Potatoes, pasta, rice
Fat	Protection and insulation (warmth) of the body	Butter, cheese, oily fish, meat
	Gives the body some energy	
Protein	Growth and repair of the body	Meat, fish, milk, eggs, cheese, lentils, soya, nuts, wheat, beans and peas
	Secondary source of energy	
Vitamin A	Helps the eyes see in dim light	Liver, eggs, butter, soft spreads, orange and yellow vegetables, e.g. carrots and apricots
	Healthy skin and tissue	
Vitamin B	Transfer and release of energy	Cereals, meat, fish, eggs, dairy products, pulses, yeast products
	Formation of red blood cells	
Vitamin C	Healthy skin	Fruit and vegetables, e.g. oranges, lemons, blackcurrants
	Helps the body heal faster and helps resist infection	
	Absorption of iron	
Vitamin D	Growth and maintenance of strong bones	Made by the body when skin is exposed to sunlight
	Absorption of calcium	Oily fish and eggs
Iron	Formation of red blood cells which carry oxygen around the body	Red meat, dark green vegetables, eggs, chocolate, dried fruit, wholegrain cereals
Calcium	Keeps bones and teeth strong	Dairy foods (milk, cheese, yogurt), white bread, canned fish, green leafy vegetables
	Healthy muscles and nerves	
Water	For all body actions	Drinking water, fruits and milk
	Removes waste products	

Non starch polysaccharide/dietary fibre

Non starch polysaccharide (NSP) is the part of food that is not digested in the body. It helps the bowels to move regularly, which reduces constipation and other bowel problems. It is needed for the digestive system to function properly. NSP may also help to lower your cholesterol level.

Different dietary needs

Some people have to follow a special diet because:

- they may need to lose weight
- they have an illness that needs to be controlled, by what they eat
- certain foods make them ill, so they have to avoid eating them.

Diabetes

Diabetics need a healthy diet and have to control sugar intake. Diabetes develops when the body cannot use glucose properly. People with diabetes need to maintain a healthy weight and eat a diet that conforms to healthy eating guidelines.

Coeliac disease

This is an intolerance to the protein gluten, which is found in wheat, barley and rye. Foods such as bread, biscuits, cakes and pasta must be avoided. Gluten-free products are available.

Calorie controlled

The general guidelines for people on a calorie controlled diet are to eat more starchy foods and cut down on fat and high sugar foods.

Nut allergies

An increasing number of people suffer from an allergy to nuts. Many food products have a statement or an allergy advice box on the label saying they contain nuts. It is not compulsory for food labels to give this type of statement.

Lactose intolerant

People with lactose intolerance cannot digest the milk sugar, lactose. Cow's milk must be avoided but cheese, yogurt and soya milk can be eaten.

Vegetarians

Vegetarians do not eat meat or fish. Vegetarians obtain protein from dairy products, nuts and pulses (beans, lentils and peas). Vegetarians need to ensure they eat a balanced diet including essential vitamins and minerals. There are different types of vegetarians.

C *Different vegetarian groups*

Group	Will not eat
Lacto vegetarian	Meat, fish, poultry, eggs
Lacto ovo vegetarian	Meat, fish, poultry
Vegan	Any animal products, e.g milk, cheese, etc.

Remember

The average adult should drink 1.8 litres of water – the equivalent of seven glasses of water per day.

Remember

Food containing NSP include: wholegrain pasta and rice, fruit and vegetables and wholemeal bread.

AQA Examiner's tip

- In the exam you may have to design products for special diets and answer questions about them.

Activity

Experiment with different types of flour to make gluten-free biscuits or cakes.

B *Vegetarian products and labelling*

Summary

You should now be able to:

explain the function and sources of different nutrients

recognise and explain the nutritional needs of different dietary groups.

2.3 Nutritional labelling

Dietary Reference Values and nutrition labelling

Nutritionists (people who study food) have a wide knowledge of the role of nutrients in health and disease. We know that people need many different nutrients if they are to maintain healthy lives and reduce the risk of diet-related diseases. The amount of each nutrient needed to stay healthy is called the nutritional requirement. These are different for each nutrient and also vary between individuals and life stages:

- women of childbearing age need more iron than men
- people who are very active will need more energy.

A *Which person needs more energy?*

Nutrient requirements vary according to: age, sex, body size and levels of activity. It is therefore very difficult to calculate specific nutrient requirements. Nutritionists have estimated the requirements for groups of people with similar characteristics such as age. These are called **Dietary Reference Values (DRVs)**. These figures cover the needs of most people in the population. Food manufacturers use these figures when adding information on food packaging.

Two DRVs which are commonly used are:

- **Reference Nutrient Intakes (RNI)** are the amount of nutrients (protein, vitamins and minerals) that are needed for almost everyone in a particular group (about 97% of the population). The level is considered to be higher than most people need.
- **Estimated Average Requirements (EAR)** are the amount of the average need for energy or a nutrient. Some people will need more, some people less.

Nutrition labelling of food products

Most manufacturers put the nutritional information on food product labels, although this is voluntary unless a special claim is made about the product, e.g. 'low in fat'.

Objectives

Understand Dietary Reference Values and why these are used.

Understand how food manufacturers present nutritional information on food labels.

Key terms

Dietary Reference Values (DRVs): scientifically calculated estimates of the amounts of nutrients needed by different groups of people.

Reference Nutrient Intake (RNI): the amount of a nutrient that is enough for most people in a group.

Estimated Average Requirement (EAR): the average need for a nutrient.

 *A food product displaying a special claim*

Food labels can be very confusing with all the different terms and symbols. Consumers have demanded a clearer system to present nutritional information.

Traffic light labelling

One system which has been developed by the Food Standards Agency is based on traffic light colours, to allow consumers to make healthier choices at a glance. This is based on the following:

- presenting separate information on fat, saturated fat, sugars and salt
- using red, amber or green colour coding to provide information on the level (high, medium or low) of nutrients in a portion of a product
- providing additional information on the levels of nutrients present in a portion of the product
- using nutritional criteria to determine the colour banding.

What do the traffic light colours mean?

High, medium or low?
Traffic lights are allocated according to criteria set by the Food Standards Agency

Per 100g criteria defined by Food Standards Agency			
	LOW Per 100g	MEDIUM Per 100g	HIGH Per 100g	HIGH Per portion
Fat	3g or less	3.0 - 20g	more than 20g	more than 21g
Saturates	1.5g or less	1.5 - 5g	more than 5g	more than 6g
Salt	0.3g or less	0.3 - 1.5g	more than 1.5g	more than 2.4g
Sugars	5g or less	5 - 15g	more than 15g	more than 18g

 Food Standards Agency Nutritional Criteria for the traffic light system

If you see a **red** light on the front of the pack, you know the food is high in something that we should be trying to cut down on. Stop and think. It is fine to have the food occasionally but we should be aware of how often we choose these foods.

If you see amber you know the food is not high or low in the nutrient.

Green means the food is low in that nutrient. The more green lights, the healthier the choice.

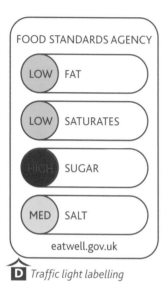

FOOD STANDARDS AGENCY

LOW	FAT
LOW	SATURATES
HIGH	SUGAR
MED	SALT

eatwell.gov.uk

D *Traffic light labelling*

SERVES 3 – THIRD PACK PROVIDES

CALS	SUGAR	FAT	SAT FAT	SALT
700	8.7g	41.7g	20.0g	2.33g
35%	10%	60%	100%	39%

OF YOUR GUIDELINE DAILY AMOUNT

A 90g slice provides...

cal 236	fat 12g
total sugars 14.9g	sat fat 7.2g
salt 0.7g	

new

V Vegetarian	Source of fibre Benefit	Each cookie contains						
		Calories	Sugar	Fat	Saturates	Salt		150g
		85	2.7g	4.3g	1.3g	0.2g		
		4%	3%	6%	7%	3%		Weight
		of your guideline daily amount						

E *Examples of 'traffic light' labels which can be seen on food products*

Many of the foods with traffic light colours will have a mixture of red, amber and greens. When you are choosing between similar products, try to go for more greens and ambers, and fewer reds, to make healthier choices.

links

Visit **www.eatwell.gov.uk/ foodlabels** for the latest information on food labels.

Nutrition panel

The traffic light colours on the front of food packs are a quick and easy guide, but if you are interested in finding out more, you will need to check the back of packs for more information. The nutrients panel gives the nutritional breakdown of more of the nutrients and includes non starch polysaccharide (dietary fibre).

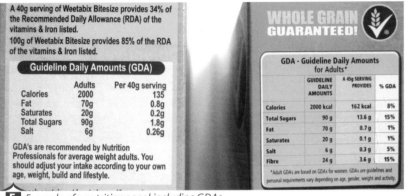

F *Example of a nutrition panel including GDAs*

Guideline Daily Amounts

Guideline Daily Amounts (GDAs) are a guide to the amounts of calories, sugar, fat, saturated fat and salt you should try not to exceed to have a healthy balanced diet. They show you what percentage of your GDA is in a portion of a product, helping you see the nutritional contribution it makes to your overall diet.

Health claims

Many foods we buy have statements on the label about their beneficial effects on the body, such as 'helps maintain a healthy heart'. This is a health claim. Rules came into effect in 2007 to help protect consumers from misleading claims, which means that any claims made about the nutritional and health benefits of a food will only be allowed if they are based on science.

Key terms

Guideline Daily Amounts (GDAs): guide to the amounts of calories, sugar, fat, saturated fat and salt you should try not to exceed to have a healthy balanced diet.

Activities

1. Collect a variety of food labels that include nutritional information from different manufacturers and retailers. Annotate the information and explain which label gives the clearer information.

2. Explain why the presentation of nutritional information may be useful to the consumer.

3. Using nutritional analysis software analyse the nutritional breakdown of a product you have made. Use the information in **C** to produce a traffic light symbol for the product.

∞ links

- Page 120 explains how to use nutritional analysis software for the controlled assessment.
- Chapter 10 includes more information about food labelling.

Summary

You should now be able to:

explain and interpret Dietary Reference Values

explain how food manufacturers present information on food labels related to nutritional information and healthy eating guidelines.

3 Combining ingredients

3.1 Raising agents

Consumers expect baked products such as bread and cakes to have a light and open texture. To create the desired texture and sufficient volume a **raising agent** is usually added.

The three most common raising agents are:

- air
- steam
- carbon dioxide.

Objectives

Understand the different raising agents and how they are used in the production of food products.

How does a raising agent work?

- The action of moisture or heat or acidity (or a combination of the three) triggers a reaction with the raising agent to produce a gas.
- Gas expands when heated.
- The gas becomes trapped as it bubbles up through the mixture.
- When heated the bubbles form a firm structure containing a network of tiny holes left by the expanded gases. This gives bread, cakes or scones a soft, sponge-like texture.

Air

Air is a mixture of gases. Air is incorporated into mixtures using mechanical methods. Air can be added in different ways:

- sieving flour – air is trapped between fine particles, e.g. cakes, batter
- creaming together fat and sugar – air becomes trapped in the mixture, e.g. cakes
- rubbing fat into flour – air is trapped as fat is rubbed into the flour, e.g. shortcrust pastry, scones, shortbread
- whisking and beating – egg white is capable of holding up to seven times its volume of air as the protein stretches, e.g. meringue, whisked cakes
- folding and rolling–air is trapped between layers, e.g. flaky pastry.

A *How does a raising agent work?*

B *Different methods to add air to a mixture*

Steam

Water turns to steam when heated at 100°C. Steam can expand up to 1600 times its original volume. Steam only works as a raising agent in mixtures that have:

- high amounts of liquid in the mixture, e.g. milk in Yorkshire pudding
- high baking temperatures which allow the liquid to quickly reach boiling point.

Steam works during baking when:

- The liquid reaches boiling point and steam is released.
- Steam forces its way up through the mixture to stretch the structure and raise it.
- The mixture bakes and sets in the risen shape.
- There are large pockets of air where the steam has escaped resulting in an open and uneven texture, e.g. Yorkshire pudding.

Steam can be combined with other raising agents, e.g. air and carbon dioxide in cakes and bread, and air in flaky pastry.

Carbon dioxide

Carbon dioxide is incorporated into mixtures in two ways, i.e:

- using chemical raising agents – such as bicarbonate of soda or baking powder
- using a biological raising agent – from the fermentation of yeast.

Carbon dioxide gas is released into the dough and when heated the gas expands and pushes up the surrounding mixture.

Chemical raising agents

These are powders that require liquid and heat to produce the carbon dioxide gas. They must be accurately measured and used in small quantities as they have strong characteristics that can affect the flavour, texture and appearance of a product.

Baking powder

Baking powder is made from a combination of alkaline and acid substances (bicarbonate of soda and cream of tartar) that react when they come into contact with moisture and warmth to produce carbon dioxide bubbles. These bubbles expand during the cooking process to cause the baked item to rise, producing a fine, delicate-textured result, e.g. in cake making.

Bicarbonate of soda

Bicarbonate of soda can be used successfully as a raising agent; however, it has a yellow appearance and a strong, unpleasant taste. It must therefore be combined with other strong flavours to disguise this, e.g. chocolate cake and gingerbread.

Biological raising agent

Yeast is a living organism that needs warmth, liquid, food and time to release carbon dioxide. This process is called fermentation and aerates the dough during bread making.

Activity

1 Fill two equal sized glasses with flour. Sieve the contents of one glass twice and carefully place back into the glass. What is the result?

2 Make up four batches of small cakes using the same quantities for each. Change the raising agent. Use: Self raising flour, plain flour, plain flour with 5g bicarbonate of soda and plain flour with 5g baking powder. Thoroughly evaluate the results assessing the effect of the raising agent on volume, density, texture, taste and appearance.

Key terms

Raising agent: increases the volume of doughs, batters and mixtures by promoting gas release (aeration).

AQA Examiner's tip

Be able to explain the reason why a raising agent is added to food products including key terms such as: gas, expands, heated, trapped, volume, texture, structure.

Summary

You should now be able to:

name the three raising agents: air, steam and carbon dioxide

explain how raising agents work in different recipes

explain the difference between a chemical and biological raising agent.

kerboodle!

Food structures

Food products are made from combining ingredients to determine appearance, texture, flavour, shape and volume. How a recipe works depends on the type and quantity of ingredients and how these react to each other. The combination of food materials using different methods of processing allows a vast range of products to be made.

Colloidal structures

Processed foods usually contain more than one ingredient. When ingredients are mixed together they usually form what is called a **colloidal structure**. Colloidal structures give texture to many products. **Colloids** are formed when one substance is dispersed through another, e.g. flour and water, but do not form a **solution**. A colloid is formed because the molecules are too big to form a solution and therefore one substance is dispersed through another. The parts of a colloidal structure may be:

- liquid, e.g. water, oil, milk
- gas, e.g. air, carbon dioxide
- solid, e.g. flour.

You will use colloidal structures many times during your making activities. The four types of colloidal structures you will need to be aware of are:

- emulsions – oily and watery liquids mixed together, e.g. mayonnaise
- foams – bubbles of gas trapped in a liquid, e.g. whisked egg white or whipped cream
- gels – a liquid held in a solid network, e.g. jam or jelly
- suspensions – solid in a liquid, e.g. white sauce, custard.

A *Different colloidal structures*

Colloidal structure	Part	Part	Example
Emulsion	Liquid, e.g. oil	Liquid, e.g. vinegar	Mayonnaise
Foam	Gas, e.g. air	Liquid, e.g. cream	Whipped cream
Solid foam	Gas, e.g. air	Liquid, e.g. baked egg white	Meringue
Gel	Liquid, e.g. water	Solid, e.g. fruit	Jam
Suspension	Solid, e.g. flour	Liquid, e.g. milk	White sauce

Gels

Gels consist of a small amount of a solid mixed in a large amount of liquid that then sets. When gels are set they are usually soft and elastic,

Emulsion

Gel

Foam

B *Examples of colloidal structures*

e.g. gelatine sets a jelly and is used to set a variety of cold desserts. If a gel is allowed to stand for a time it may start to 'weep'. This loss of liquid is known as **syneresis.**

Suspensions

A suspension is a solid held in a liquid, e.g. the starch grains in flour are mixed into a liquid when a sauce is made. The mixture has to be stirred to keep the solids evenly mixed in the liquid or the solid will sink to the bottom resulting in a lumpy sauce.

Foams

A mixture of gas and liquid is called foam. The trapped air bubbles increase the volume of the mixture. The foam creates an aerated product with a smooth texture, e.g. whipped cream.

A solid foam is formed if the mixture is cooked, e.g meringue. Heating coagulates (sets) the protein producing a solid fine-mesh structure.

Emulsions

A mixture of two liquids is called an emulsion. When water and oil are mixed together, they form an emulsion but they do not mix permanently. If left to stand the oil will rise to form a separate layer on top of the water, e.g. French dressing.

To stop the liquids separating an **emulsifying agent** must be added to create what is called a **stable emulsion**. Mayonnaise is a stable emulsion of oil and vinegar, when egg yolk (lecithin) is used as an emulsifying agent. Manufacturers have to use emulsifiers in creamed cake mixtures, margarine, chocolate and salad cream.

Activity

1   Carry out a comparative shop of the different ingredients added to salad dressings. Evaluate the results. Design and make an interesting salad dressing to add to a pasta salad.

2   To gain making experience using colloidal structures make:

lemon meringue pie, lemon/chocolate mousse, sponge flan and layered desserts.

Summary

You should now be able to:

explain the term colloidal structure

explain the difference between: suspensions, gels, foams and emulsions

recognise colloidal structures in the products you make.

Remember

Syneresis also occurs when protein foods are overheated. The protein coagulates (sets) and squeezes out fat and water, e.g melted cheese and overcooked scrambled eggs.

C *Example of a solid foam: meringue*

Key terms

Colloidal structure: when two substances are mixed together.

Colloids: formed when one substance is dispersed through another.

Solution: when a solid dissolves in a liquid, e.g. salt in water.

Syneresis: loss of a liquid from a gel on standing, e.g. custard.

Emulsifying agent: a substance that will allow two immiscible liquids (substances that do not mix) to be held together, e.g. lecithin in egg yolk.

links

See Chapter 8 Additives.

3.3 Cake-making methods

There are four basic methods of combining and mixing ingredients when you make a cake. The ratio and type of ingredients are different for each method. The methods are:

- rubbing-in
- creaming – traditional and all-in-one
- melting
- whisking.

A *Basic ingredients required for cake making*

The main ingredients in cake making are fat, sugar, eggs and flour. All methods use a raising agent and often a liquid such as milk is also added. Additional ingredients, e.g. flavourings can be added to give the cake particular characteristics.

The functions of ingredients

Flour:
- Forms the **structure** of the cake
- As the cake is heated, protein (gluten) in the flour **sets** the framework and shape
- **Dextrinisation** occurs, starch converts into a sugar. On heating the sugar caramelises resulting in a **golden** surface

Fat:
- Adds **colour** and **flavour**
- Holds air bubbles (foam) which creates texture and **volume**
- Produces a short crumb or rich even **texture** dependent on the ratio of fat and method used
- Increases the **shelf life**

Sugar:
- **Sweetens** and adds **flavour**
- When creamed with fat, helps to hold **air** in the mixture
- Caramelisation gives **colour**

Raising agents:
- **Aerates** the mixture increasing **volume** and resulting in a light **texture**

Eggs:
- **Trap** air when whisked into a foam
- **Coagulate** (set) on heating
- **Emulsify** – holds the fat in emulsion and keeps it stable
- Add colour, flavour and nutritional value

B *Function of ingredients*

c *Cake making methods*

Cake making method	Ratio	Main ingredients				Full list of ingredients	Raising agent	Method/procedure	Characteristics
		Flour	Fat	Sugar	Egg				
Rubbing-in method *Examples* • Rasberry buns • Rock cakes • Fruit loaf	8:4:4:1	200g	100g	100g	1 (x50g)	200g Plain flour 100g Fat 100g Caster sugar 1 Egg (50g) 10ml Baking powder 30ml Milk Flavourings 175g Dried fruits	*Chemical* Baking powder or self raising flour *Mechanical method:* Adding air: Sieving Rubbing-in	• Rub fat into flour • Add additional ingredients • Add liquid • Bind together • Knead gently • Shape • Bake 200°C/Gas 6	• Produces dry and open cakes with a short shelf life • Well-risen with a rough surface
Creaming method Traditional or All-in-one *Examples* • Victoria Sandwich • Maderia cake • Small buns	1:1:1:1 Equal	100g	100g	100g	2 (2x50g)	100g Self raising flour 100g Soft spread or butter 100g Sugar 2 Eggs (100g) *All-in-one.* *5ml baking powder may be added to aid rising.*	*Chemical* Self raising flour includes baking powder *Mechanical method:* Adding air: Sieving, Creaming fat and sugar	• Fat and sugar creamed together • Eggs added slowly, beating well • Gently fold in the flour • Add to container • Bake 190°C/Gas 5 *All-in-one.* All ingredients creamed together using a hand or electric whisk	• Produces fluffy and light sponge • Even and fine texture • Longer shelf life due to ratio of fat
Melting method *Examples* • Gingerbread • Parkin • Brownies • Flapjack	8:4:8:2	200g	100g	50g + 150g black treacle	1 (1x50g)	200g Plain flour 100g Fat 50g Soft brown sugar 150g Black treacle 50g Golden syrup 125ml Milk 1 Egg (50g) 5ml Bicarbonate of soda 10ml Ground ginger 5ml Mixed Spice	*Chemical:* Bicarbonate of soda (Spices mask the flavour and colour from the raising agent)	• Fat is melted with the treacle, syrup and sugar • Stir in dry ingredients • Add eggs and milk • Pour into container • Bake 180°C/Gas 4	• Soft, moist, sticky and even texture • Flavour develops during keeping • Long-shelf life
Whisked method *Examples:* • Swiss roll • Sponge flan • Gateaux	1:1:2	50g	None	50g	2 (2x50g)	50g plain flour 50g caster sugar 2 eggs (100g)	Steam from the eggs *Mechanical method:* Adding air: Sieving Whisking	• Eggs and sugar whisked together until mixture has doubled in volume • Flour gently folded into the mixture • Pour into container • Bake 200°C/Gas 6	• Very light sponge • Even and soft texture • Short shelf-life due to no fat

The function of ingredients

When designing and making cakes it is very important to understand the function (job) of each ingredient.

Faults in cake making

When testing and experimenting in the test kitchen or making products for the first time the results are not always perfect. As a food technologist it is important to recognise the reasons why and then correct the working errors. To be able to do this you need to understand the ingredients and processes you are using.

D *Common faults and causes*

Fault	Cause
Peaked cracked top	Oven too hot
	Too much mixture for size of tin
	Baked on too high a shelf in oven
	Too stiff or too wet a mixture
	Over mixing cake batter
Cake sinks	Too much sugar causing collapse of the structure
	Too much raising agent
	Undercooking, caused by wrong temperature and time
	Disturbed during cooking causing structure to collapse
Sugary speckled crust	Too much sugar
	Wrong type of sugar used
	Insufficient creaming
Close heavy texture	Too much liquid in the mixture
	Insufficient raising agent used
	The creamed mixture has **curdled** and does not hold sufficient air
	Whisking method
	Eggs and sugar not beaten enough
	Over beating when adding flour
Coarse and open texture	Too much raising agent used
	Insufficient mixing of flour
Cake very dry	Overcooking of the cake
	Insufficient liquid used
	Too much raising agent
Fruit has sunk	Too much liquid to carry the weight of fruit
	Too much sugar and raising agent

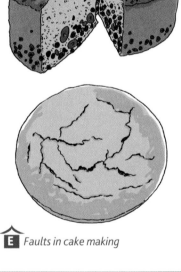

E *Faults in cake making*

Large scale cake production

Cakes are enhanced by the addition of attractively applied finishes such as buttercream, glacé icing, fondant icing, marzipan, royal icing, crystallised fruits or other decorations.

There have been many technological advances in commercial cake production. There are many celebration/novelty cakes available to buy.

F *Celebration cake*

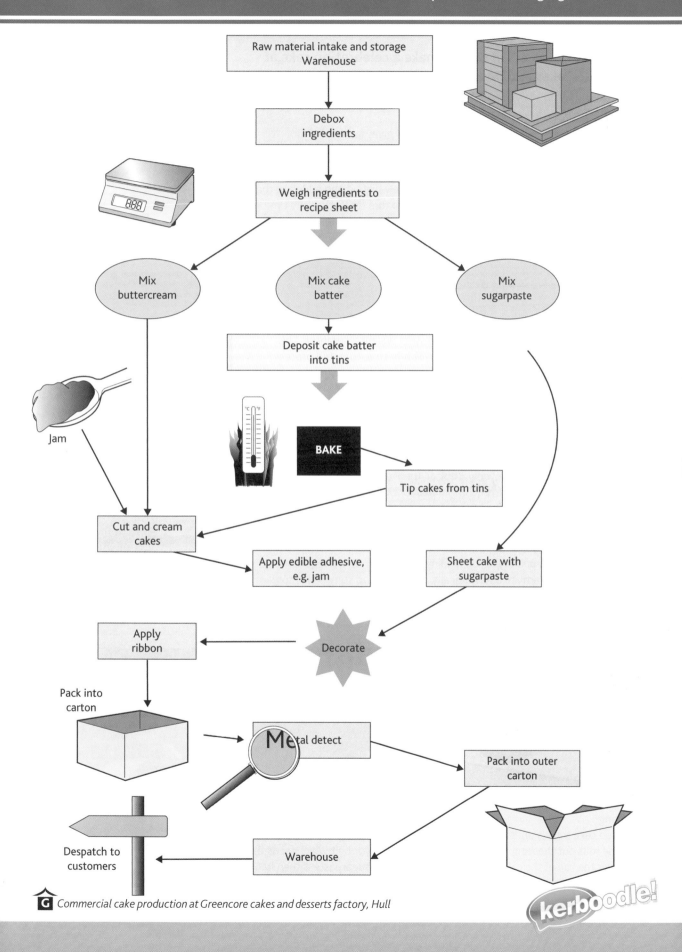

Raw material intake and storage Warehouse

Debox ingredients

Weigh ingredients to recipe sheet

Mix buttercream

Mix cake batter

Mix sugarpaste

Deposit cake batter into tins

Jam

BAKE

Tip cakes from tins

Cut and cream cakes

Apply edible adhesive, e.g. jam

Sheet cake with sugarpaste

Apply ribbon

Decorate

Pack into carton

Metal detect

Pack into outer carton

Despatch to customers

Warehouse

Commercial cake production at Greencore cakes and desserts factory, Hull

kerboodle!

Step-by-step process of how to make a celebration cake at Greencore Cakes and Desserts Factory, Hull.

Weighing
All ingredients are weighed and batch coded. This gives full traceability throughout the manufacturing process.

Mixing
Bulk ingredients such as flour, sugar and water are weighed and added to the mixer. The mixers can mix up to 300kg of cake mixture (batter) at any time – if the ingredients were bought from a supermarket, this would equate to 200 packs of butter, 75 bags of flour, 75 bags of sugar, 21 litres of water and 1200 eggs! Specialised ingredients, which help make the cake easier to process, are added. This helps the cake last longer – usually about 4 weeks rather than a home made cake which would last about 3 days.

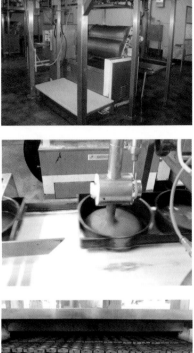

Depositing
Cake batter is deposited into cake tins which have been automatically sprayed with grease. The cake batter is deposited to strict weight tolerances. This ensures consistency in the finished product. This is an automated process which uses a moving conveyer belt which feeds the cake tins straight into the oven.

Baking
Large travelling ovens (about half the length of a football pitch) are used. These can hold hundreds of cakes at a time. The travelling ovens have three zones to allow the cake to bake differently at each stage of the cooking. This makes sure the cakes all rise evenly, cook all the way through and brown evenly. Bake times can vary from 50 minutes to 2 ½ hours. The cakes are then cooled under strict humidity and temperature control and then stored for up to 48 hours before being processed further.

Cutting & creaming
The cake is cut into 2 or 3 layers, depending on the customer's requirements. Jam and/or buttercream are automatically deposited and sometimes spread by hand. Weight checks are taken here and at every step to ensure product consistency and quality. The cakes are then put onto cakeboards so they are easier to lift in and out of the packaging when they are finished.

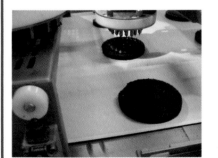

Sheeting
The filled cake travels under a sugarpaste (soft fondant icing) sheeter. This machine rolls out the icing and feeds it onto the conveyor belt which feeds the icing over the top of the cake which covers it fully. The sugarpaste is then hand moulded around the cake and trimmed, ready for decorating.

Decorating

Cake decorations are made or bought in already made (standard components). These are all hand placed to a specified design. The decorating team are trained to a high standard. If chocolate curls are applied to the sides of a chocolate cake, these can be done either by hand or automatically.

Finishing off

For a finishing touch, some cakes have a coloured ribbon or a plastic or cardboard collar placed around the edge.

Completed product

The finished products are checked against quality standards. Again, this ensures the consumers are satisfied with the finished products. The cakes are then placed into boxes which are then sealed and labelled with a 'Best Before' date before being put into large boxes or plastic crates for shipping to the supermarkets to be sold.

⬭links

See 3.1 Raising agents on page 28.

Summary

You should now be able to:

explain the different cake-making methods

explain the functions of ingredients in cakes

identify the faults in cake making and suggest remedies to the problem

successfully make recipes using different cake-making methods

recognise how large scale cake manufacturers produce a consistent outcome

decorate a cake to achieve a quality outcome.

Activities

1. Adapt a traditional Victoria sandwich cake and produce designs to show how to improve/develop the:
   a nutritional value
   b flavour and texture
   c appearance and finish.

2. Produce a revision aid to help learn the function of ingredients in cakes.

3. Compare the making of a Victoria sandwich cake in a test kitchen with large scale production. Produce a detailed comparison table.

3.4 Pastry-making methods

Pastry-making methods

Pastry is basically a dough, made from flour, fat, salt and water that is then rolled out and used as a base, cover or envelope for sweet or savoury fillings. The wide range of pastries made today vary in texture and taste according to the proportion of fat used, the way in which it is incorporated with the flour and the method used to shape the dough. Each type of pastry has its own distinctive appearance, texture and flavour. There are many different types of pastry:

- short crust
- rough puff
- choux
- flaky
- suet
- sweet short crust
- filo
- hot water crust
- sweet shortcrust.

It is important to be familiar with the different types of pastry; however for now you only need to know how to make short crust and rough puff pastry.

Rolling and folding rough puff pastry

Rough puff pastry is created by rolling and folding the dough resulting in a light and layered pastry.

- A dough of flour, small cubes of fat, water and lemon juice is made and rolled to a rectangle.
- This is then folded and re-rolled several times until there are several layers.
- Rolling and folding allows air to be trapped in the layers.
- The edges need to be sealed to prevent the air escaping.
- The pastry is cooked at a high temperature which causes the fat to melt.
- The fat is absorbed by the flour leaving air spaces.
- The air then expands when heated and the water in the mixture converts to steam which expands. This creates the pastry layers.

Objectives

Understand the different pastry-making methods and the ratio of ingredients for each method.

Understand the functions of ingredients in shortcrust and rough puff pastry.

Understand the faults that occur when making pastry.

Understand the nutritional differences of pastry.

Remember

Remember it is important to demonstrate your making skills for the controlled assessment. This could be done by carrying out experiments and investigations using pastry-making methods. You could experiment making other types of pastry, e.g. choux pastry.

Remember

Remember shortcrust pastry has a crumbly texture; this is explained on page 16.

A *Pastry making method*

Pastry making method	Ratio	Ingredients		Full list of ingredients	Method/procedure	Raising agent	Characteristics
		Flour	Fat				
Shortcrust pastry *Examples:* • Pies • Jam tarts • Quiche • Pasties	2:1	200g	100g	• 200g Plain flour • 100g Hard / white fat • 1g Salt • 40ml Cold water Rich shortcrust pastry includes 50g of sugar and I egg yolk.	• Fat is rubbed into the flour to look like fine breadcrumbs • Water gradually added to make a soft dough • Knead lightly • Chill • Roll out • Bake: 190°C/Gas 5	**Air:** • Sieving the flour Air trapped during rubbing in.	• Short-crumb and light texture • Crisp texture when baked
Rough puff pastry/Flaky *Examples:* Sausage rolls Turnovers Eccles cakes	4:3	200g	150g	• 200g Plain (strong) flour • 150g Fat (mixture of white fat with butter or hard fat) • 1g Salt • 10ml Lemon juice • 100ml Cold water	• Rub fat into flour • Mix together water and lemon juice • Add to mixture and mix to a dough • Roll the dough • Fold the dough into three • Repeat four times. • Chill • Shape • Bake: 200°C/Gas 6 **Flaky pastry:** Only a quarter of the fat is rubbed in and the remaining fat is added to the layers during folding.	**Air:** • Trapped between the layers as the dough is folded and rolled. **Steam:** • Steam created between the layers from the fat due to high cooking temperature	• Layered crisp flakes • Short and crisp texture.
Choux pastry *Examples:* Profiteroles Eclairs	3:1	75g	25g	• 75g Plain flour • 25g Butter or soft spread • 2 Eggs (100g) • 125ml Water	• Fat and water melted in a saucepan • Flour added to form a thick sauce • Cool slightly, gradually beat in eggs. • Pipe • Bake: 220°C/Gas 7	**Air:** • Beating the eggs **Steam:** High water content.	• Light and airy • Hollow structure • Crisp texture

The function of ingredients

Ingredients have different functions depending on the type of pastry that is made. The ratio of each ingredient determines the end result in terms of texture, taste and finish.

Functions of ingredients: pastry

Flour
- Soft plain flour (low gluten content) is used in shortcrust pastry to give a short crumb.
- Strong plain flour (high gluten content) used in flaky/rough puff pastry to give the pastry elasticity.
- Flour forms the structure of the pastry.

Fat
- In shortcrust pastry the fat coats the flour granules resulting in a crumbly texture.
- Fat traps air between the layers in flaky/rough puff pastry.
- Adds colour and flavour.

Water
- Binds the dry ingredients together.
- Lemon juice is added to the water when making flaky pastry /rough puff to strengthen the gluten required for rolling and stretching.

Salt
- Adds flavour.
- Stengthens gluten.

B *Function of ingredients*

C *Basic ingredients required for pastry making*

Faults in pastry making

Making pastry is a skilled process. It is important to be able to recognise any faults that could occur during the making process. Consider how pastry is made by food manufacturers and the possible faults that could occur. Quality control procedures are essential throughout the making process.

links

See example of pastry samples on page 122.

D *Common faults and causes in pastry making*

Fault	Cause
Shortcrust pastry	
Pastry is hard and has a tough texture	Over kneading and heavy handling by hand or machine
	Incorrect proportions of ingredients – designated tolerances not followed. Too much water added
	Incorrect oven temperature – too cool
Pastry is blistered	Oven set at too high a temperature
	Fats insufficiently mixed into flour
	Uneven addition of water
Pastry is fragile and crumbly	Too much fat
	Insufficient water
	Over mixing the fat into flour
Pastry has shrunk during cooking	Over working of the pastry during kneading and rolling
Rough puff/flaky	
Pastry has not flaked well	Oven temperature too cool. Steam has not been produced quick enough
	Insufficient liquid added, the mixture was too dry
	Pastry folded and rolled unevenly
	Pastry has not rested sufficiently in a cool environment
	Pastry folded too thinly
Shrinkage	Dough not relaxed enough after rolling and make-up

Activity

1 To gain making experience, make the following:

   a Short crust pastry: jam tarts

   b Rough puff: cheese and onion pasties.

Remember

When making a pastry case it is sometimes necessary to cook the pastry to prevent the bottom becoming soggy and undercooked. This is called 'baking blind'. The pastry is first baked with a lining weighted with something such as dried beans or baking beans.

Top tips for pastry making

- The kitchen, utensils and ingredients should be cool. Pastry doesn't like warmth as the fat starts to melt and therefore does not trap as much air.

- The measurements in the recipe must be adhered to as any deviation will change the outcome.

- Handle the dough mixture as little as possible, working as quickly and lightly as you can. This minimises development of the flour's **gluten** content otherwise the pastry may become too elastic, difficult to roll, inclined to shrink, and tough in texture.

- Add the liquid a little at a time. Too much liquid creates a tough dough, whilst too little gives a crumbly result.

- Pastry seems to be better when it is left to chill. Once the dough is formed, chill for 30 minutes, this helps relax the gluten and set the fat, making the dough manageable and less likely to shrink.

Key terms

Gluten: protein found in flour.

Modifying a recipe

Pastry and many pastry products contain high proportions of fat; they should not be eaten regularly as part of a balanced diet.

Our food range is just part of at Weight Watchers has to offer. e also run thousands of friendly etings across the UK every week.

To find out more, call

08457 123 000

or visit

weightwatchers.co.uk

010076 726402

E

NUTRITION INFORMATION

Typical Values	Per 100g	Per 165g
Energy	824 kJ	1359 kJ
	197 kcal	324 kcal
Protein	7.0g	11.6g
Carbohydrate	21.2g	35.0g
of which sugars	3.1g	5.1g
Fat	9.3g	15.4g
of which saturates	4.6g	7.6g
Fibre	1.6g	2.6g
Sodium	0.21g	0.35g
Salt	0.5g	0.9g

WEIGHT WATCHERS

WEIGHT WATCHERS Cheese & Onion Quiche can help slimming or weight control only as part of a calorie controlled diet. For details of Weight Watchers Meetings contact: Weight Watchers (UK) Ltd. Tel 08457 123000 Visit our website: www.weightwatchers.co.uk

GUIDELINE DAILY AMOUNTS

Use the following table as a daily guideline

Each Day	Women	Men
Calories	2000	2500
Fat	70g	95g
Salt	6g	6g

One 165g serving provides 324 Calories, 15.4g of Fat and approx 0.9g of Salt

WEIGHT WATCHERS on foods and beverages is the registered trademark of WW Foods, LLC and is used under license.

WEIGHT WATCHERS for services and *POINTS* are the registered trademarks of Weight Watchers International, Inc. and are used under license. All rights reserved.

Manufactured by Greencore Prepared Foods under license of Weight Watchers International, Inc.

If you are not entirely satisfied with this product, please write to Greencore Prepared Foods, Sheffield S26 5PF, quoting the use by date and stating when and where purchased. This does not affect your statutory rights.

HEATING GUIDELINES

May be eaten warm or cold.
For best eating quality serve warm as below. All appliances vary, the following are guidelines only.
Ensure the Cheese & Onion Quiche is piping hot before serving.
If using a fan-assisted oven, reduce time and temperature to suit.
To oven heat from chilled:
Preheat oven to 200°C, 400°F, Gas Mark 6. Remove all packaging including foil. Place quiche onto a baking tray and heat in the middle of the oven for 15-20 minutes. Stand for 5 minutes before serving.

INGREDIENTS

Skimmed Milk, Wheat Flour, Onion (14%), Fromage Frais, Vegetable Oil, Egg (5%), Egg White, Maize Flour, Cheese (with Colour: Beta-Carotene) (2.5%), Bulking Agent: Polydextrose; Modified Maize Starch, Red Leicester Cheese (with Colour: Annatto) (1.5%); Double Cream (0.8%), Chive, Salt, Flavouring, Mustard Powder, Flour Treatment Agent: L-Cysteine Hydrochloride; White Pepper, Stabiliser: Xanthan Gum.

ALLERGY ADVICE

Contains: milk, wheat, gluten, egg, mustard.

FREEZING GUIDELINES

This product has been frozen and returned to chill temperature. Further freezing will not affect the quality. Freeze on day of purchase and use within one month. Please ensure product is fully defrosted before reheating. Do not refreeze after defrosting.

STORAGE INSTRUCTIONS

Keep refrigerated.

Activity

2 Look at the two food product labels, **E** and **F**. Produce a report explaining:

a which you consider to be is the healthiest; justify your answer.

b how the manufacturer has reduced the fat content for product **E**.

c how product **F** could be developed to increase the originality and sensory appeal.

d the function of ten of the ingredients in product **F**.

e how you would develop the pastry and filling to produce an original product? Make the product and carry out a nutritional and sensory evaluation.

Remember

Remember you cannot make pastry with low fat spread due to the high water content. The ratio of fat to flour must also remain unchanged to achieve an accurate result.

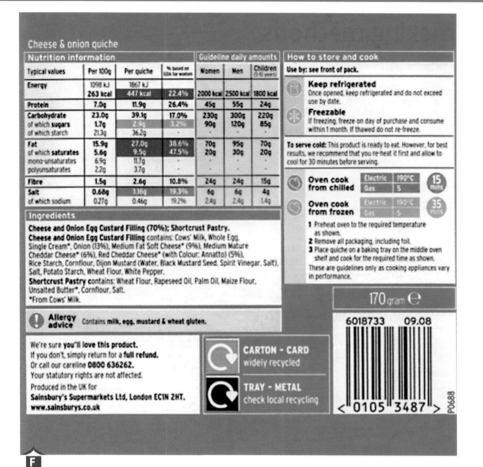

Cheese & onion quiche

Nutrition information

Typical values	Per 100g	Per quiche	% based on GDA for women	Women	Men	Children (5-10 years)
Energy	1098 kJ	1867 kJ				
	263 kcal	447 kcal	22.4%	2000 kcal	2500 kcal	1800 kcal
Protein	7.0g	11.9g	26.4%	45g	55g	24g
Carbohydrate	23.0g	39.1g	17.0%	230g	300g	220g
of which sugars	1.7g	2.9g	3.2%	90g	120g	85g
of which starch	21.3g	36.2g				
Fat	15.9g	27.0g	38.6%	70g	95g	70g
of which **saturates**	5.6g	9.5g	47.5%	20g	30g	20g
mono-unsaturates	6.9g	11.7g	-			
polyunsaturates	2.2g	3.7g	-			
Fibre	1.5g	2.6g	10.8%	24g	24g	15g
Salt	0.68g	1.16g	19.3%	6g	6g	4g
of which sodium	0.27g	0.46g	19.2%	2.4g	2.4g	1.4g

Ingredients

Cheese and Onion Egg Custard Filling (70%); Shortcrust Pastry.
Cheese and Onion Egg Custard Filling contains: Cows' Milk, Whole Egg, Single Cream*, Onion (13%), Medium Fat Soft Cheese* (9%), Medium Mature Cheddar Cheese* (6%), Red Cheddar Cheese* (with Colour: Annatto) (5%), Rice Starch, Cornflour, Dijon Mustard (Water, Black Mustard Seed, Spirit Vinegar, Salt), Salt, Potato Starch, Wheat Flour, White Pepper.
Shortcrust Pastry contains: Wheat Flour, Rapeseed Oil, Palm Oil, Maize Flour, Unsalted Butter*, Cornflour, Salt.
*From Cows' Milk.

! Allergy advice Contains milk, egg, mustard & wheat gluten.

We're sure **you'll love this product.**
If you don't, simply return for a **full refund.**
Or call our careline **0800 636262.**
Your statutory rights are not affected.
Produced in the UK for
Sainsbury's Supermarkets Ltd, London EC1N 2HT.
www.sainsburys.co.uk

CARTON – CARD
widely recycled

TRAY – METAL
check local recycling

How to store and cook

Use by: see front of pack.

Keep refrigerated
Once opened, keep refrigerated and do not exceed use by date.

Freezable
If freezing, freeze on day of purchase and consume within 1 month. If thawed do not re-freeze.

To serve cold: This product is ready to eat. However, for best results, we recommend that you re-heat it first and allow to cool for 30 minutes before serving.

Oven cook from chilled | Electric | 190°C | **15 mins**
| Gas | 5 |

Oven cook from frozen | Electric | 190°C | **35 mins**
| Gas | 5 |

1 Preheat oven to the required temperature as shown.
2 Remove all packaging, including foil.
3 Place quiche on a baking tray on the middle oven shelf and cook for the required time as shown.
These are guidelines only as cooking appliances vary in performance.

170 gram e

6018733   09.08

< 0105 3487 >

P0688

F

Activity

3 In groups make a batch of rough puff/short crust pastry and use a packet mix of the same pastry. Make sausage rolls using the two batches of pastry. Produce a full evaluation comparing: methods, cost, timing and sensory characteristics.

Summary

You should now be able to:

name different types of pastry

explain the two different pastry-making methods

explain the functions of ingredients in short crust and rough puff pastry

identify the faults in pastry making and suggest remedies to the problem

successfully make recipes using different pastry-making methods

analyse two existing recipes and suggest nutritional and sensory developments.

AQA Examiner's tip

- Learn the basic ratio and ingredients for standard pastry recipes. You may be asked for quantities when answering a design question.

- Know why pastry may not be successful and try to relate this to making pastry in a factory.

Using sauces

A sauce is a flavoured liquid, such as milk, thickened to a suitable consistency. A sauce is used in the preparation of recipes or as a dressing, coating or filling. The basic structure of most sauces is made up of four components:

A *Sauce components*

Component	Example: cheese sauce	Example: sweet and sour sauce
Liquid	Milk	Pineapple juice and water
Thickening agent	Plain flour	Cornflour
Seasoning	Salt	Soy sauce
Flavouring	Cheese	Tomato/vinegar/sugar

Objectives

Understand the uses of a sauce in food production.

Understand the different methods to make a sauce.

Understand the functions of ingredients in different sauces.

Understand the differences between making a sauce in a test kitchen compared to a manufacturing environment.

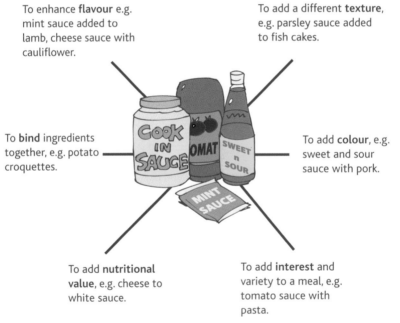

To enhance **flavour** e.g. mint sauce added to lamb, cheese sauce with cauliflower.

To add a different **texture**, e.g. parsley sauce added to fish cakes.

To **bind** ingredients together, e.g. potato croquettes.

To add **colour**, e.g. sweet and sour sauce with pork.

To add **nutritional value**, e.g. cheese to white sauce.

To add **interest** and variety to a meal, e.g. tomato sauce with pasta.

B *Reasons for using sauces*

Types of sauces

There are many different types and flavours of sauce. The characteristic of a sauce is determined by the method by which it is made, the proportion and type of ingredients used.

- **Starch based** – starch from wheatflour, arrowroot and cornflour is used to thicken a liquid. When the mixtures are heated and stirred they will thicken. This process is known as gelatinisation, e.g. cornflour to make custard, wheatflour to make cheese sauce and arrowroot to glaze a flan. These sauces are smooth and may be **bland** and therefore need to have flavourings added.

Remember

A liquid that has been thickened is considered to be a sauce; therefore custard, gravy, fruit **couli** are all types of sauces.

links

Gelatinisation is explained in more detail in Chapter 1.

Emulsion is explained more in 3.2 Food structures on page 30.

- **Fruit or vegetable sauces** – cooked or raw fruit or vegetables can be puréed to produce a smooth sauce. This is usually done using a liquidiser and passing through a sieve, e.g. tomato sauce and strawberry purée. These sauces are vibrant in colour rather than rich and creamy.

- **Egg-based sauces** – when eggs are heated they coagulate and thicken a sauce, e.g. egg custard. These sauces are rich, thick and creamy.

- **Oil/water emulsion**s – if oil and water are thoroughly mixed, they become dispersed in each other and form an emulsion. Egg yolk stabilises the emulsion and forms the sauce, e.g. mayonnaise. These sauces are usually rich, glossy and smooth.

- **Cream** – cream can be mixed with flavourings, e.g mustard, peppercorns, and heated. The liquid evaporates and thickens the sauce. These sauces are rich.

Starch-based sauces

A starch-based sauce, using a flour to thicken, can be made by the:

- blended method
- roux method
- all-in-one method.

> **Remember**
>
> Sauces can be flavoured in many different ways, using a variety of herbs, spices, vegetables, cheeses, etc. As part of your controlled assessment, testing the flavouring and thickness of sauces could feature as part of your experimental and investigation work.

> **Key terms**
>
> **Couli:** fruit that has been puréed, sieved and then thickened.
>
> **Bland:** lack of flavour/taste.

C *Starch-based sauces*

Method	Ingredients	Features	Making/process
Roux sauce Examples: ▪ Cheese sauce ▪ Parsley sauce	Flour/ cornflour Soft spread/ butter Liquid: milk, stock	▪ Thickness depends on the amount of flour and liquid ▪ Equal amounts of fat and flour ▪ Used to pour, coat and bind ▪ Traditional method	▪ Melt the fat and stir in the flour. Cook for one minute. Take off the heat ▪ Add the liquid gradually, stirring continuously ▪ Heat slowly to thicken, stirring all the time
All-in-one sauce Examples: ▪ Cheese sauce ▪ Parsley sauce	Flour/ cornflour Soft spread/ butter Liquid: milk, stock	▪ Same ingredients / proportions as the roux method ▪ Quick and easy method ▪ Use a whisk to avoid lumps	▪ Add liquid, flour and fat to a saucepan ▪ Heat slowly whisking all the time
Blended sauces Examples: ▪ Custard ▪ Sweet and sour sauce	Cornflour/ arrowroot Liquid: fruit juice	▪ Simple sauce to make ▪ Pouring consistency	▪ Blend starch with a small amount of liquid ▪ Heat remainder of the liquid ▪ Stir liquid into blended starch ▪ Reheat, stirring all the time until boiling point is reached

D *Making a roux sauce*

Consistency/viscosity of a sauce

The proportion of flour and fat to the amount of water will determine the thickness of the sauce. We call this the **viscosity**. Different recipes may involve making sauces to different thicknesses, e.g. a dessert would be served with a pouring sauce but a pasta dish will require a thicker sauce to coat the pasta. The three categories of thickness include:

- pouring
- coating
- binding.

E *Basic white sauce proportions*

Sauce	Ratio	Liquid	Flour	Fat
Pouring	16:1:1	250ml	15g	15g
Coating	10:1:1	250ml	25g	25g
Binding	5:1:1	250ml	50g	50g

The function of ingredients

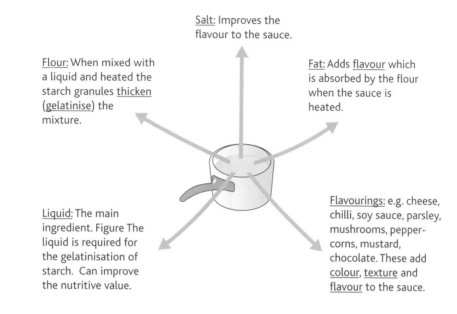

Salt: Improves the flavour to the sauce.

Flour: When mixed with a liquid and heated the starch granules <u>thicken</u> (<u>gelatinise</u>) the mixture.

Fat: Adds <u>flavour</u> which is absorbed by the flour when the sauce is heated.

Liquid: The main ingredient. Figure The liquid is required for the gelatinisation of starch. Can improve the nutritive value.

Flavourings: e.g. cheese, chilli, soy sauce, parsley, mushrooms, pepper-corns, mustard, chocolate. These add <u>colour</u>, <u>texture</u> and <u>flavour</u> to the sauce.

F *Function of ingredients in a sauce*

Faults when sauce making

When making a sauce it is essential that the making process is followed accurately. The table explains some of the common faults when sauce making.

G *Common faults and causes when sauce making*

Fault	Cause
The sauce is too thick	Inaccurate weighing of thickening agent, e.g. cornflour
	Inaccurate measuring of the liquid, e.g. milk
The sauce is lumpy	Insufficient agitation. The mixture was not continually stirred and therefore the starch granules have sunk to the bottom of the pan
	Liquid added too quickly when making a roux sauce
	Extra flour added to the mixture that was not blended with a liquid before adding
	Heated too quickly and vigorously
The sauce is a poor colour	The flour granules have burnt at the bottom of the saucepan
	Metal spoon used in a metal saucepan resulting in a grey coloured sauce
The sauce is bland/poor flavour	Insufficient flavouring or seasoning added
	Mild cheese used
	Floury taste – insufficient cooking of the fat and flour when making a roux sauce

H *Commercial sauces*

> **Remember**
>
> If extra thickening agent is required to achieve the correct consistency it must be first blended with a liquid and then stirred into the sauce. If the thickening agent is added without a liquid this will result in a lumpy sauce.

Commercial sauce making

Food manufacturers have developed many different sauces. We can buy many sauces in jars off the supermarket shelves; these are called ambient sauces and are stored at room temperature. There has been a significant increase in chilled sauces that need to be stored between 0–5°C. Buying a ready-made sauce, a standard component, allows for a quick meal to be produced. The case study below shows how a sauce is manufactured on a large scale.

A day in the life of a cooked chilled sauce

1 The ingredients for the sauce are weighed and placed in collection containers ready for use.

2 The cooking process is controlled by a computer program (Computer Aided Manufacture).

3 When all the ingredients have been loaded into the cooking vessel, the sauce is cooked. This is the first Critical Point (CCP).

4 The hot sauce is pumped into the high risk department where it is deposited into pots. Further checks are then carried out, including temperature checks, metal detection and date coding.

5 The pots of sauce are then chilled to below 5°C. This is the final CCP in the process. The pots are then placed into outer packaging.

6 The sauce is then stored in a chilled warehouse to await distribution to supermarkets.

Activities

1. Read the case study related to sauce production. Explain the main differences between how a sauce is made in a test kitchen compared to commercial production of a sauce.

2. Prepare a cheese sauce for macaroni cheese. In groups compare making the sauce by the all-in-one method and the roux method.

3. Design and make a sauce that could be added to pasta.

AQA Examiner's tip

Practical experience is a good way to remember recipes, methods, problems which can occur and how to rectify them.

Summary

You should now be able to:

name different types of sauce

explain the functions of ingredients in different sauces

identify the faults when making a sauce

explain the difference between making a sauce in a test kitchen and large scale manufacturing.

3.6　Bread

" *There is something magical about making bread – no other recipe gives you so much in return from such basic ingredients, which literally change in front of you. Making bread is a true kitchen pleasure that everyone should learn.* "

Barney Desmazery, BBC Good Food, October 2008

Objectives

Understand the functions of ingredients in bread making.

Understand the stages of bread making.

Understand the faults that can occur during bread making.

Bread is a **staple food** in many of the world's cultures. There has been an increase in the types of bread sold in supermarkets and bakeries in the UK. There are a number of reasons for this including our multicultural society and increased foreign travel. Types of bread include: Italian ciabatta, Indian naan, French baguettes, Irish soda bread and Middle Eastern pitta bread. Breads can be made more interesting by having different ingredients added, e.g. basil, cheese, sultanas, coconut, etc, they can also be shaped in different ways and include a variety of toppings and finishes.

Bread is a good source of carbohydrates, protein, B-group vitamins and the minerals calcium and iron. Wholemeal bread is also a very good source of NSP (dietary fibre). Therefore bread is very nutritious.

A *Types of bread and finishes*

Function of ingredients in bread

Strong plain flour (normally white)

■ Has a high **gluten** content. Gluten is a protein in the flour; when mixed with water it, forms an elastic and stretchy texture. Elasticity of the gluten enables doughs containing yeast to stretch and hold carbon dioxide gas in small pockets creating a light and open texture. Gluten sets when cooked at high temperatures and forms the framework and shape of the bread.

Yeast

■ Yeast is the raising agent used in bread making.
■ Produces carbon dioxide gas, which makes the bread rise – this is called **fermentation**.
■ Yeast needs: food, moisture, warmth and time in order to grow and ferment.

Liquid

■ Binds the dry ingredients together and helps in the development of gluten.
■ The liquid should be lukewarm (25°C to 35°C) to aid yeast fermentation. If too hot the yeast will be destroyed, if too cold the action of yeast is slowed down.

Salt

■ Adds flavour.
■ Controls the action of yeast.
■ Strengthens the gluten.

Remember

Wholemeal flour can also be used. For best results half wholemeal flour to half white strong flour should be used.

Remember

There are three types of yeast used in bread-making:

■ fresh yeast, which is a firm, moist, cream-coloured block available from bakeries
■ dried yeast, which comes in small granules that are first reconstituted with warm water and sugar
■ powdered (or 'easy-blend' or 'fast-action') dried yeast which is sprinkled straight into a bowl of flour.

Fat

- Enhances colour and flavour.
- Increases shelf-life, prevents the bread going stale.

Sugar

- Small amounts aid fermentation.
- Can be added to sweet or rich yeast doughs.

Mixing and kneading
Flour, salt and fat are mixed with the yeast and water. Flexible dough formed. Kneaded to stretch the dough and develop the gluten and form an elastic dough.

Fermentation
The dough is left to stand to rise. This is called proving. The yeast produces carbon dioxide gas, which causes the dough to rise.

Knocking back
To create an evenly textured bread the dough is kneaded to release some of the gas. It is left to rise again.

Shaping
The gluten should know be thoroughly distributed. The dough is shaped and left to prove again.

Baking
The heat sets the gluten and stops the yeast working. The heat sets the shape.

 Bread-making stages

> ## Key terms
>
> **Staple food**: a food that forms the basis of a traditional diet – wheat, barley, rye, maize, or rice, or starchy root vegetables such as potatoes.
>
> **Fermentation**: when yeast produces carbon dioxide.
>
> **Gluten**: protein found in flour.

Bread making

Faults when bread making

C *Faults when bread making*

Fault	Cause
Loaf has not risen well; has a heavy and closed texture	Incorrect flour used, low gluten content
	Too much salt
	Yeast killed before the loaf is baked
	Insufficient kneading or proving
	Dough has been over fermented, resulting in a breakdown of the gas pockets in the dough
	Insufficient liquid resulting in dough which is too stiff to allow expansion
Uneven texture with large holes	Dough not 'knocked back' properly
	Dough left uncovered during rising
Dough collapses when put into oven	Over-proving
Crust breaks away from the loaf	Under-proving
	Dough surface dried out during proving
	Oven too hot

Remember

The amount of salt added to the mixture needs to be carefully weighed. The addition of too much salt will kill the yeast. Direct contact between the yeast and salt must be avoided.

Remember

A bread maker can be used in a test kitchen. Bread makers are automated and mix and cook the bread in the same machine. A bread maker uses CAM (Computer Aided Manufacture). The electronics control all the processes and timings. A bread maker can be programmed to turn on in the middle of the night so that fresh bread is waiting for you in the morning.

Commercial bread making

Nine million loaves are consumed every day. To produce this number of loaves bakeries produce bread over a 24-hour period. The stages are the same as making bread in the test kitchen but extra ingredients (bread improvers, vitamin C) are added to speed up the process.

links

To see different methods of commercial bread production go to **www.warburtons.co.uk**.

Case study

How bread is made in a Warburtons bakery

Step 1
Flour arrives in tankers at the bakery. A computer-controlled mixer weighs flour and water. Yeast, salt and other ingredients are added automatically. Batches of dough are mixed every few minutes.

Step 2
Batches of dough are divided into portions for 400g or 800g loaves at a speed of 125 loaves per minute. The dough is passed through a spinning machine. The bread is kneaded in huge machines.

Step 3
The dough is passed on a conveyor belt to the next stage of production. The bread making process is controlled by Computer Aided Manufacture (CAM). Conical moulders shape dough into balls to produce 8,000 loaves per hour. First prover allows dough to 'rest' for 6–8 minutes.

Step 4
The dough is developed by rolling and cut into equal size pieces. An elastic dough is required to hold carbon dioxide gas in small pockets to create a light and open texture.

Step 5

Four pieces of dough are added to the bread pans. This results in even rising and consistent texture. Dough spends 50 minutes in the final prover and expands in controlled humidity and temperature.

Step 6

Bread travels through the oven for 20–25 minutes. Lids are added to some tins to produce flat-topped bread. 6,000 large or 8, 000 small loaves are baked per hour.

Step 7

Strict quality controls take place to ensure every Warburtons loaf is the same. The bread is sliced and regular samples are taken away to be inspected to ensure that Warburtons' high quality is maintained.

Step 8

Bread is packaged, date marked, stacked and put onto lorries for distribution the same day it is made.

Activities

1   Evaluate a range of bread products commenting on: shaping, finishing techniques, flavourings, ingredients and sensory characteristics.

2   Investigate the gluten content of self-raising, strong plain and wholemeal flour. Mix 30g flour with water. Place the mixture in the centre of a 'J' cloth and fasten. Run under cold water until the liquid is clear. Bake on a tray at 220°C. Evaluate the results.

3   Explain the differences between making bread in a test kitchen and commercial bread production.

AQA *Examiner's tip*

Understand the functions of ingredients in bread and the effect of additional ingredients on a yeast mixture, e.g. addition of sugar, eggs in a sweet dough.

Summary

You should now be able to:

explain the functions of ingredients in bread

explain the stages in bread making

successfully make a range of bread products.

Acids and alkalis

All foods are made up of a variety of natural food chemicals. The type and amount of food chemicals present in the food create the conditions:

- acid
- alkali
- neutral.

Acids and alkalis influence the changes that occur when some ingredients are combined. Acids and alkalis are found in many different foods and they have many uses in food production. Both acids and alkalis have an effect on the flavour, appearance, texture and nutritional value of food products.

It is often useful to know how acidic or alkaline a food is. Acidic foods usually taste sour, e.g. lemon juice. Sodium bicarbonate is an alkali; this has an unpleasant taste and needs to be mixed with a stronger flavouring when used in cooking, e.g. ginger to make parkin.

Objectives

Understand which ingredients are acid and alkali.

Understand the effect of acids and alkalis on the texture, appearance, flavour and storage of food.

Understand how acid and alkali ingredients can be used in food production.

Colour	red		orange		yellow		green		blue		navy blue			purple
pH	1	2	3	4	5	6	7	8	9	10	11	12	13	14

increasingly acidic ⟵   neutral   ⟶ increasingly alkaline

A *pH scale*

Acids and alkalis are measured using the pH scale. In science lessons you will have tested how acidic or alkaline particular substances are using Universal Indicator paper or a probe. A value of 7 on the pH scale means the substance is neutral, a value of 1–6 indicates an acid and a value of 8–14 an alkali.

Acids

It is important to know the acidic level of foods because they can affect the end result when making a food product. From the activity you will know that fresh fruits are acidic. If fruits are mixed with milk products they can cause the mixture to **curdle**. If a food technologist is aware of this they can take steps to prevent this from happening. The use of acids has many beneficial effects in food production both in the test kitchen and commercially.

Activities

1. Test the pH of the following foods using Universal Indicator paper. If the food is solid it should be dissolved in a little water before testing.

   Banana, lime juice, sugar, milk, bicarbonate of soda, fizzy drink, orange juice, baking powder, salt, apple.

2. Draw a results table and conclude the activity.

Uses of acids

Uses of citric acid (lemon juice)

When some cut fruits, e.g. sliced apple and banana, have contact with oxygen in the air a reaction takes place resulting in the fruit turning brown. This is called **enzymic browning**. Citric acid, in the form of lemon juice, is used to prevent enzymic browning, e.g. when making a fruit salad. This improves the appearance of the food product. Other uses of citric acid are:

- to set (coagulate) some chilled desserts. The protein in cream and condensed milk sets when mixed with citric acid, e.g. in cheesecake
- to help set jam. The acid helps to form a gel.

Uses of acetic acid (vinegar)

- Meat can sometimes be tough. If the meat is marinated in a vinegar solution this tenderises the meat and results in a softer texture. Meats are often marinated before adding to a barbecue.
- Acetic acid added to meringue results in a soft marshmallow texture, e.g. pavlova.
- It adds flavour, e.g. in salad dressings. A marinade is also used to improve the flavour of food.
- It is used to preserve food products, e.g. onions, beetroot, cabbage. Acid conditions prevent the growth of microorganisms.

Uses of ascorbic acid (vitamin C)

- Ascorbic acid is added during commercial bread production to speed up the fermentation process – the release of carbon dioxide.
- Other acids are used, for example, in cheese production to coagulate milk and to make the raising agent baking powder.

Use of alkalis

Alkalis are used less in food production. Their main use is as a **raising agent**. Bicarbonate of soda is an alkali and produces carbon dioxide when heated, e.g. when making parkin and gingerbread. The cracked appearance on biscuits such as ginger nuts is due to the addition of a small amount of bicarbonate of soda.

B Enzymic browning

C Uses of vinegar

∞links

See more example of raising agents in Chapter 3.

Activity

3 Make soft cheese by adding citric acid to form curds (solid lumps) and whey:

  a Heat 300ml of full cream milk to 37°C.

  b Add the juice of a lemon.

  c Line a sieve with muslin and place over a bowl.

  d Pour the mixture through the muslin and leave to drain. The solid left is the curd and a type of soft cheese.

  e Experiment adding flavourings to the curd cheese.

Key terms

Curdling: when a mixture separates and becomes lumpy.

Enzymic browning: reaction between a food product and oxygen resulting in a brown colour, e.g. sliced potato has brown patches when sliced and left in the air.

Summary

You should now be able to:

explain which ingredients are acid and alkali

explain the uses of acids and alkalis in food production.

kerboodle!

5.1 Standard components

■ What are standard components?

A **standard component** is a pre-prepared item/ingredient that is used in the production of another product.

In the food industry manufacturers buy standard components from other food manufacturers.

Example of standard components include:

- pizza bases
- ready-to-roll icing
- stock cubes
- sauces for pasta
- ready-made pastry
- fruit fillings for pies
- marzipan
- cake decorations
- grated cheese
- dried herbs.

Using a standard component ensures:

- a standard shape of a product, e.g. buying a pastry case to make a strawberry flan
- a standard size or weight, e.g. using a pizza base
- a standard flavour, e.g. using a stock cube to add to soup
- accurate proportion and ratio, e.g. packet sauce mix to make a cheese sauce.

■ Why do food manufacturers use standard components?

As well as saving time and sometimes costs, using a standard component can ensure a **consistent** product is produced. If a manufacturer is to use a standard component from another supplier it is essential that a precise and accurate **specification** is produced. Before deciding to use a standard component the manufacturer will check:

- whether the product uses any additives, e.g. artificial colours and flavourings
- the nutritional profile of the product, e.g. salt and saturated fat content
- the source of ingredients, e.g. air miles, use of organic ingredients
- sensory qualities, e.g. taste, texture and appearance.

It is important that the component meets the standards set by the manufacturer.

Advantages of using standard components

There are many advantages of using standard components in the test kitchen and for manufacturers.

 Examples of standard components

- Ensures consistency of flavour, texture, weight, shape and colour. Allows for an exact replication of the product each time it is made.
- Saves preparation time.
- The quality is guaranteed if an accurate specification is produced.
- Can speed up the manufacturing process.
- Less effort and skill required by staff, e.g. a less skilled workforce required.
- Less machinery and specialist equipment required. Complex production lines take up lots of space and expensive to buy mass production machinery.
- Less risk of cross contamination. High risk products can be prepared elsewhere.
- Components can be bought in bulk.
- Can save the manufacture money.
- Specific requirements of companies can be met.
- A wider range of products can be produced, from the same standard component, e.g. pizza bases.

Disadvantages of standard components

As well as the many advantages of using standard components there are some disadvantages.

- Depending on another manufacturer to supply a product can be less reliable, e.g. a sandwich manufacturer relies on a company to supply bread. There is a power cut, the bread cannot be cooked, the sandwiches cannot be made and there is a delay in the production. This could cost the company a lot of money.
- The components can be more expensive than manufacturing yourself.
- The sensory qualities of the components may not be as good.
- A considerable amount of storage space may be required.
- Special storage conditions may be required.
- Time required for ordering and supplying.
- The quality may not be as good as you wish.

The use of standard components in the test kitchen

When making food products in the test kitchen it will be necessary to use standard components for some of the reasons mentioned above. When carrying out experiments and investigations it is acceptable to make use of some ready made ingredients. When you are producing the prototype for the controlled assessment you will need to make each part of the prototype yourself. Reliance on standard components can affect your making mark.

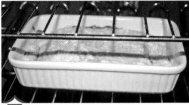

B *Cooking a pasta product in the test kitchen*

Activities

1. Produce a list of standard components that you could use to make: pizza, lasagne, chocolate gateau, cheesecake and spring rolls.

2. Make two different pasta products: one using fresh ingredients and other using standard components. Complete a sensory, cost and nutritional comparison.

Key terms

Standard component: a pre-prepared item used in food production.

Consistent: the same quality each time a product is made.

Specification: a document that contains all the details of a final product.

Summary

You should now be able to:

define the term standard component

give examples of standard components used in the test kitchen and in food manufacturing

explain the advantages and disadvantages of using standard components.

kerboodle!

Examination-style questions

 Questions will require you to show knowledge and understanding of the functions, working characteristics and the processing techniques when designing and making food products.

AQA Examiner's tip Rarely are the exam q's based on one area, they often include more than one section of the spec.

1. Many manufacturers are keen to choose ingredients that will make their products healthy options. The table below compares different types of flours used in bread products.

| Type of flour | Per 100g | | | |
	Kcal	Fat	Fibre	Iron
Wholemeal SR	310 kcal	2.2g	8.6g	3.9mg
White strong	341 kcal	1.4g	3.7g	2.1mg
Brown plain	323 kcal	1.8g	7.0g	3.2mg

Using information from the table:

(a) (i) Which type of flour is most suitable for use in a healthy option product?
Give reasons for your choice *(3 marks)*

(ii) Which type of flour is most suitable for use in bread making?
Give reasons for your choice. *(5 marks)*

(b) (i) List four main ingredients needed to make a basic white loaf of bread.
Give a different reason for use for each ingredient. *(8 marks)*

(ii) Explain the nutritional properties of these ingredients. *(4 marks)*

2. The table below shows food facts for two similar products.

PRODUCT A		PRODUCT B	
Wholemeal short crust pastry made with polyunsaturated fat, reduced fat cheese, fresh sliced tomato, onions and herb topping		Flaky pastry made with butter, topped with full fat cream cheese, onion flavouring and sundried tomatoes	
Energy	146 kcal	Energy	229kcal
Protein	1.7g	Protein	1.5g
Carbohydrates	23.8g	Carbohydrates	65g
Fat	4.8g	Fat	11g
Fibre	0.7g	Fibre	0.5g

(a) (i) Explain which product provides the healthier option. *(6 marks)*

(ii) Why are ratios and proportions important when making pastry? *(2 marks)*

(iii) What are the effects of changing the polyunsaturated fat to butter? *(2 marks)*

(b) (i) Name two different ways flaky pastry is sold as a standard component. *(2 marks)*

(ii) What are the advantages and disadvantages of using standard components? *(6 marks)*

3   A local manufacturer is developing a fish in sauce product. Using the roux method they will use flour, liquid, butter and a flavouring.

(a)   Name a main ingredient used to

(i) thicken the sauce

(ii) flavour the sauce

(iii) provide a glossy appearance to the sauce. *(3 marks)*

(b) (i) Describe two different ways the manufacturer makes sure that the sauce does not become lumpy. *(4 marks)*

(ii) Explain what is meant by **gelatinisation.** *(2 marks)*

AQA
Examiner's tip

These questions are based on general food products but similar questions may be asked on any chosen food product. You may like to practise similar questions for the food product of your choice.

Processes and manufacture

In this section you will find out about the ways food can be produced using a range of processes and equipment that are carried out in the test kitchen and when mass produced. It is essential all food products made are safe to eat and this happens by adding certain ingredients or processing foods using different techniques.

To be able to produce good quality food products every time it is vital to use the correct equipment and processes. This will often involve using electrical equipment and basic cooking skills. You will also find out that once a food product has been made, correct food storage is essential to keep it at its best.

■ What will you study in this section?

After completing Processes and manufacture (Chapters 6–10) you should have a good understanding of:

- the use and effect of additives in food production
- safe storage of food and food products
- how the use of certain pieces of equipment can ensure the consistent production of high quality foods
- the differences and requirements of large scale production
- new technological developments in food production
- the importance of correct labelling and packaging of foods.

■ Making/cooking

It is great fun to learn about the process and manufacture of foods by carrying out various cooking activities and investigations. These activities will help you understand some of the reasons why food has to be processed. You can carry out investigations and making activities to see:

- how emulsifiers and flavourings are used in food production
- how the use of cutters get uniformity of size and shape when making biscuits
- how steaming is a healthy way to cook green vegetables
- how to stop cross contamination occurring when cooking high risk foods
- how small pieces of equipment help ensure consistency of making every time
- how new technologies produce brand new products.

How will you use this information?

You will be tested on this knowledge in the examination. It is also important that you apply the knowledge to ensure consistently high quality products in your controlled assignment. This usually comes about by using electrical and mechanical equipment during making and product development. You will also need to consider food safety and hygiene in the controlled assessment.

6 Equipment in the test kitchen

6.1 Types of equipment

Hand-held, mechanical and electrical equipment are used in the development of prototypes in the test kitchen to ensure a consistently high quality product every time. They also save time, energy, ease the making process and ensure uniformity of results. All these pieces of equipment need to be used safely and **hygienically.**

The processes using electrical equipment could be carried out by hand-held equipment, e.g. grating cheese using a grater; however, a food processor carries out the same task in a small proportion of the time producing identical results every time.

Which processes can be carried out using electrical equipment?

A *Many processes can be carried out using electrical equipment*

- weighing
- mixing
- chopping
- slicing
- kneading

- whisking
- liquidising
- shredding
- beating
- freezing

Using pieces of equipment safely and effectively

- Before plugging in electrical equipment make sure it has been assembled correctly.
- Use electrical equipment away from a water source.
- Avoid using electrical equipment with wet hands.
- Make sure that the correct attachments are used to carry out the process required, e.g. grating blade on a food processor to prepare carrots for coleslaw.
- Make sure the lid is correctly closed before turning on the machine, e.g. of a liquidiser before liquidising soup.
- Make sure the process has fully finished before removing the foods.

B *Make sure the lid is correctly closed*

How do we ensure that food is hygienically produced using equipment?

- The food technologist working in the test kitchen has high standards of personal hygiene, e.g. correct clothing, clean hands, no jewellery.
- All equipment is dismantled and each piece cleaned after use to avoid **cross contamination.**
- All equipment is stored in a clean area.

How do we ensure uniformity of food products?

Using electrical equipment provides a consistently high quality product every time. The use of smaller equipment such as cutters, cake tins, moulds and muffin cases ensures a uniformity of finished product.

C *A way of ensuring uniformity*

Remember

- Hygienic use of equipment means food spoilage will not occur.
- Safe use of equipment means an accident will not occur.

Activities

1   Which piece of electrical equipment will best carry out each of the following tasks:

   a   liquidising soup
   b   rubbing-in shortcrust pastry
   c   slicing peppers for a stir fry
   d   whisking meringues for pavlova.

2   Make two coleslaws, one using a food processor and one using a sharp knife and grater. Compare your results.

Summary

You should now be able to:

understand the advantages of using electrical equipment to carry out food preparation processes

list four rules to use a food mixer safely and hygienically

use three different small pieces of equipment to make high quality products.

Uses of equipment

When making a **prototype** food product the equipment used in a test kitchen is chosen to carry out the processes efficiently and effectively.

What task does each piece of equipment carry out?

Electronic scales measure accurately to as little as 0.1g. This accuracy is vital in product development. As products are being developed small changes to the prototype ingredients may be made to get the flavour, texture and consistency that are required. Flavourings, such as chilli powder for a curry, may need adapting very slightly to get the correct 'spiciness'.

Food processors have a range of attachments that enable them to carry out many different tasks. They grate, slice, grind, shred, dice, purée, mix, knead and whisk. As the tasks are carried out it is possible to time the processes to ensure a **consistent** product every time. The attachments that carry out grating, slicing, etc. produce a consistent sized product every time, e.g. the size of grated carrot pieces.

Mixers have a range of attachments so that different mixing processes can be carried out, e.g. whisking, mixing, kneading. These enable different consistencies and textures to be achieved in the food products including the trapping of air to make a well-risen product. The speed and length of the mixing can also be controlled to ensure a consistent final product, e.g. egg whites whisked for meringue.

Liquidisers and hand held blenders have a very specific process to carry out which is making solid ingredients into liquid. Using different lengths of processing time can vary the texture and consistency of the food products, e.g. puréed fruit for a smoothie.

Bread makers and ice cream makers have a very specific process to carry out and are only involved in the production of one type of food product. Once the sequence of making has been programmed in to the maker the machine completes its task and a finished product results.

How to ensure a consistent product every time
- The same process is carried out over and over again.
- The speed of the process can be altered.
- The timing of each process is exact.
- The length of time the process is carried out can be altered.
- The attachments can be changed to get a different result altogether.

Large scale equipment used in food manufacture

Lots of the processes carried out in the test kitchen are scaled up in the factory production of the same food products. Large scale equipment ensures a consistently high quality result every time and is often monitored by computers to ensure exact weights, processing times and temperatures.

Objectives

Understand which piece of equipment is best used to carry out a process successfully and efficiently.

Understand how the use of equipment achieves consistency and quality every time.

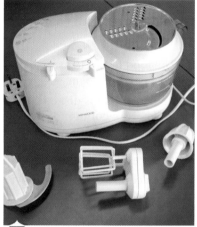

A *Food processor*

B *Ice cream maker and bread maker*

C *Large scale equipment used in food manufacture*

Tunnel ovens are used in the production of small identical items such as biscuits and involve a conveyor belt passing through a heated tunnel to cook the food product. The amount of time it takes the conveyor belt to pass through the oven is the length of cooking time the food needs.

Depositors are huge tubes which fill containers such as pastry cases; they measure an exact amount in the container every time, e.g. cake mixture in to a pastry case for Bakewell tart.

A **mandolin** is used to slice and cut foods evenly every time.

Floor standing mixers work like a huge food processor and mix large quantities of ingredients consistently each time for a specified amount of time.

An **enrober** coats a product with another ingredient to give it an outer layer; again this is usually placed above a conveyor belt that carries the product to be enrobed.

Activities

1   Using a hand blender, food processor and liquidiser make three fruit smoothies and compare the consistency of the finished smoothies. Comment on your results.

2   Explain the function of each of the attachments of a food processor.

AQA Examiner's tip

- Understand why using electrical equipment gives a consistently high quality product every time.
- Know the processes that electrical equipment can be used for in food production.

Summary

You should now be able to:

list the tasks a food processor can be used for in food production

explain why a blender gives a good consistency when making soup

explain some of the equipment used in large scale production.

Key terms

Prototype: the first version of a product that is being developed.

Consistent: the same every time.

The choice of cooking equipment and cooking method affects the nutritional quality of the food product. Water-soluble vitamins B and C are often lost if cooked in water using methods such as boiling, as they dissolve into the water and the water is generally thrown away. Vegetables, a good source of vitamins, are frequently boiled. By using other methods of cooking, such as steaming, the water-soluble vitamins remain in the vegetables and are therefore more nutritious. When using an electric steamer the food does not come into contact with the boiling water as the water is retained in the base of the unit; the vegetables are cooked in the steam as it rises.

Objectives

Understand how the correct selection of cooking equipment can result in a healthier product.

Understand how cooking methods affect the nutritional quality of foods.

A Vitamins B and C often escape when food is boiled

Microwave ovens

Microwave ovens also ensure that water-soluble vitamins remain in foods. Microwaves penetrate the food and the water and fat molecules in the food absorb their energy. These molecules **vibrate** which causes heat that cooks the food. As microwave cooking times are short and very little if any water is added for cooking, the vitamin B and C content of vegetables cooked in this way remains high.

Key terms

Vibrate: to move up and down very quickly.

Appliance: a piece of electrical equipment.

Electric griddles

Too much fat in the diet can be harmful to health and by choosing a dry method of cooking such as grilling it is possible to reduce the amount of fat in a food. Electric griddles cook the food by heating ridged hot plates one above the food and one beneath. This allows the fat to run out of the food and it is collected underneath the **appliance**. Barbecue cooking works by the same principle. Other foods that traditionally may have had fat added during the cooking process, e.g. corn on the cob, can also be cooked on these grills without the need to add fats or oils therefore making them healthier.

Nutritional content

Table **C** compares the fat, vitamin C and NSP (fibre) content of potatoes when a range of cooking methods and equipment are used.

B *Electric griddle*

C *Compare the nutritional content of potatoes cooked in a variety of ways*

Method of cooking	Amount of fat per 100g	Amount of fibre per 100g	Amount of Vitamin C per 100g
Potato crisps	37.6g	10.7g	27mg
Boiled potatoes	0.1g	1.4g	6mg
Chipped potatoes	6.7g	3.0g	9mg
Jacket potatoes	0.2g	3.0g	14mg

Activities

1. Using food tables compare the vitamin C content of 100g boiled cabbage and 100g steamed cabbage, compare the fat content of 100g of fried bacon and 100g of grilled bacon.

2. Make a steamed treacle sponge and compare it to a treacle sponge made in the microwave.

3. Cook a vegetable such as broccoli in three different ways and compare the finished result for taste, colour and texture.

AQA Examiner's tip

- Understand how water soluble vitamins are lost during cooking.
- Understand how selecting different cooking methods can result in a lower fat product.

Summary

You should now be able to:

explain why microwave cooking is a suitable method of cooking to preserve water soluble vitamins

explain the advantages of cooking on a electric grill if you are on a low fat diet.

7 Food safety and hygiene

7.1 Food spoilage

Foods cannot be stored for long periods of time without changes occurring to the taste, texture and colour of the food. **Microorganisms** and enzymes cause food spoilage and more seriously can cause food poisoning. The three types of microorganisms causing food spoilage are bacteria, yeasts and moulds. Microorganisms exist all around us, in the soil and air, on animals and humans and on equipment and preparation surfaces if poor hygiene occurs, but they can not be seen other than through a microscope. Positively microorganisms are used in the food industry to produce food products such as cheese, yogurt, bread and quorn.

What conditions do bacteria need to grow?

- Temperature: rapid bacterial growth occurs at 37°C; however growth occurs between the temperature 5°C to 63°C – this is known as the danger zone.
- Foods: High risk foods (those containing large amounts of protein and water) are the best medium for bacterial growth.
- Time: in the correct conditions bacteria reproduce by dividing in two and this can occur as often as once every 10 to 20 minutes meaning that within a few hours one bacterium becomes several million.
- Moisture: most foods contain moisture and so are ideal for maximum bacterial growth.

How does food poisoning happen?

Food poisoning occurs if harmful microorganisms **contaminate** food and are then allowed to grow. It is difficult to know if bacteria are present in food as they do not affect the appearance, taste or smell of the food. Bacteria that cause food poisoning are known as **pathogenic** bacteria, e.g. clostridium botulinum. These are very harmful and may cause death in young children and the elderly. Poor hygiene during the storage, preparation or serving of food can also result in food poisoning.

The main types of food poisoning bacteria are:

- Salmonella: the most common form of food poisoning found in Britain. Its **symptoms**, such as diarrhoea, vomiting, headaches and stomach pains, appear 12 to 36 hours after eating the affected food. Main sources of the bacteria are eggs and poultry.
- Campylobacter: has symptoms of diarrhoea and headaches that occur 1 to 11 days after eating the food. Main sources of the bacteria are meat, poultry and shellfish.
- Staphylococcus aureus: a bacteria that creates toxins which cause food poisoning. They are present in the nose, throat and skin of humans and so can be transferred to any foods by poor personal hygiene. The symptoms that appear one to six hours after eating

A *Microorganisms affect foods – red pepper and cheese*

the food are vomiting, abdominal pain and diarrhoea. Main food sources are meat, poultry and salads.

How do we stop bacteria growing in food?

If we alter any of the optimum conditions needed by bacteria for growth it will stop them from reproducing or greatly slow down the rate of reproduction.

Temperature

What are the critical temperatures affecting bacterial growth in foods?

By looking at the thermometer it can be seen that:

−18°C	bacteria are dormant and not able to reproduce
0°C to 5°C	bacteria are 'sleeping' and reproduce extremely slowly
5°C to 63°C	bacteria reproduce most actively – this is known as the danger zone
37°C	optimum temperature for bacteria to reproduce
72°C	bacteria start to be destroyed and are not able to reproduce.

Food Bacteria grow best on high risk foods, foods that have a high protein and water content, e.g. meat, eggs, dairy produce, fish and cooked rice all provide ideal conditions for growth.

Time Food needs to be prepared as quickly as possible and should not be allowed to wait around in danger zone temperatures before storing. Therefore chilling the food after preparation is important.

Moisture Care must be taken when preparing foods with a high moisture content.

How does preservation help ensure a longer shelf life?

If we alter any of the optimum conditions needed by bacteria for growth it will stop them from reproducing or greatly slow down the rate of reproduction. The main methods of preservation are:

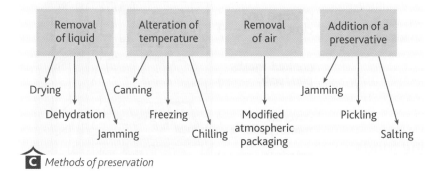

C *Methods of preservation*

Other substances can also cause food spoilage

Physical and chemical spoilage of foods can occur during their storage, preparation and cooking. The main physical contaminates are jewellery, flakes of nail varnish, metal from equipment, hair and insects. Chemical contaminates are most likely to be cleaning fluids.

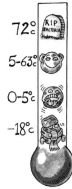

B *Critical temperatures affecting bacteria*

Activity

Which would be the best method of preservation for the following foods: eggs, cooked lasagne, chicken curry, bacon, chicken and vegetable soup?

Summary

You should now be able to:

explain how bacteria reproduce and the effect this can have on foods

recognise the critical storage temperatures and why these are important

list five ways of preserving foods

name different food poisoning bacteria and it's symptoms.

kerboodle!

Freezing and chilling are the most popular forms of extending the storage life of foods as the colour, taste, texture, flavour and nutritional value of the food is altered least of all the food preservation methods. These foods are readily available in the freezer and chill cabinets of supermarkets.

How do freezing and chilling affect food products?

Freezing **preserves** food for between one week and up to one year depending on the food. Most foods can be frozen easily; exceptions are those with a high water content, e.g. strawberries and salad ingredients, or those with a **colloidal structure**, e.g. sauces, which may separate on thawing. Commercially frozen food is rapidly frozen using blast, plate or **cryogenic freezing** methods. This ensures small ice crystals form in the food cells and that no damage is caused to the structure of the food. Slow freezing results in large ice crystals being formed that damage the food cell walls. On thawing these foods have changes to their flavour, texture and nutritional values.

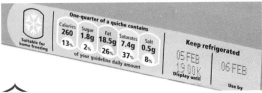

A   *A food label has to show how to store foods safely*

Fast freezing involves reducing the core temperature of the food from 0°C to –18°C in 12 minutes. Freezing stops microorganisms from growing; however, they are not destroyed but are dormant. Domestic freezers are usually at a temperature of –18°C but commercial freezers for longer-term storage are kept at –29°C.

Chilling does not preserve food but extends its **shelf life** for a few days. There is very little change to the taste, texture, flavour and nutritional value of these foods; they are almost identical to fresh produce. Chilled foods are stored between 0°C and 5°C as this prevents the growth of Listeria monocytogenes, a common food-poisoning bacteria.

Foods that are to be sold as chilled foods are prepared using the cook chill method. This involves preparing and cooking the food and then it is blast chilled to a temperature of below 5°C in 90 minutes. The foods must be stored at less than 4°C for a maximum of five days. At home these foods are stored in a refrigerator.

Modern lifestyles mean that more consumers then ever have less time to prepare food and therefore buy foods that just need reheating.

Essential rules for reheating foods

- Ensure frozen foods are fully defrosted.
- Follow the reheating instructions on commercially prepared food products.
- All foods need to be reheated to above 72°C.
- Ensure that the core temperature of high risk foods and dense foods are at 72°C by using a food probe.

cheese and dairy →

cooked meats pies/paté →

covered raw meats/poultry →

salads in lidded boxes →

B   *Ensure no cross contamination occurs when storing foods in the refrigerator*

How to use a temperature probe

- Plug in the probe.
- Ensure it is clean (wipe using an antibacterial wipe).
- Check that the digital reading is set to 0.
- Place into the centre of the food to be tested.
- Hold in place for two minutes.
- Ensure the digital reading is static and take the reading.
- Remove from the food, read again, clean using the antibacterial wipes and reset.

C *Using a food probe*

AQA *Examiner's tip*

- Understand that freezing preserves food but chilling is a short-term method of extending shelf life.
- Understand how and why a temperature probe is used to ensure correct reheating of high risk and dense foods.

⚭ links

Modified starch prevents the separation of sauces on thawing, see 1.1 Starch on page 10.

Activities

1 Make macaroni cheese and freeze a small portion and chill a small portion of the food. Defrost and reheat after three days. Compare the taste, texture, appearance of the frozen and chilled macaroni cheese. Has the quality of the macaroni been affected?

2 Compare the packaging of four different main meal food products. Which symbols indicate that the food can be frozen and what methods can be used to reheat the foods?

Key terms

Preserve: to keep food fit to eat.

Colloidal structure: when two substances are mixed together.

Cryogenic freezing: food is immersed or sprayed with liquid nitrogen.

Shelf life: the length of time a food product can be kept and be safe to eat.

Summary

You should now be able to:

understand the differences and similarities between freezing and chilling foods

explain how food should be correctly stored in a refrigerator

explain the reasons why correct reheating of food will stop food poisoning.

All people who work in the food industry need a good knowledge and understanding of how food is produced safely. The key starting point is a high level of personal hygiene.

What is good personal food hygiene?

- Wear clean protective clothing, e.g. overalls, hairnet and/or or hat, shoes.
- Tie back hair and cover with a hair net or hat.
- Clean hands (using **antibacterial** wash and hot water) before touching food.
- Cover all cuts with a blue waterproof dressing.
- Keep fingernails short and scrubbed without nail varnish.
- Remove all jewellery, e.g. rings, watches, earrings.
- Wear disposable gloves (usually blue in colour) if possible before touching food.
- Do not chew or smoke near food.
- Do not touch your ears, nose or hair when handling food.
- Report any illness, e.g. sickness, diarrhoea, cold or flu to a supervisor (this is a legal requirement).

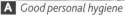
A *Good personal hygiene*

What is good kitchen hygiene?

- Keep different food preparation areas for different foods, e.g. high risk – fish and meat, medium risk – vegetables and meat pies, low risk – potatoes and jam. This prevents cross contamination.
- The use of colour coded equipment, e.g. knives and chopping boards, to prevent cross contamination. This equipment must only be used for the specified foods and cleaned immediately after use.

Red:	raw meat
Blue:	fish
Green:	fruit and vegetables
Yellow:	cooked meats
White:	dairy products

- Good facilities for cleaning and waste disposal.
- Equipment and work areas kept in good condition, e.g. stainless steel sinks, regular removal of waste.
- Methods for preventing insect contamination, e.g. fly screens.

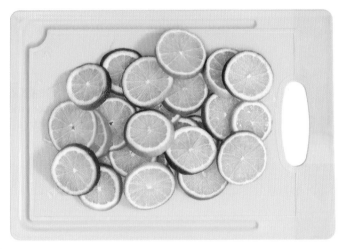

B *Green chopping boards are used for fruit and vegetables.*

After buying a food product, what are the risks to food safety?

When food is produced in a factory a production run occurs. At each stage of the making of the product checks are in place to ensure the finished product is of good quality and is safe to eat. These checks are quality control checks and hazard checks.

Activities

1 In pairs make a high risk product: one of the pair makes the product whilst the other draws up a safety and hygiene checklist and scores their partner.

2 Imagine you have purchased a lasagne from the chilled cabinet of a supermarket at midday and intend to eat it in the evening. Where might possible food poisoning hazards occur?

AQA Examiner's tip

Understand how some equipment is used to prevent cross contamination of foods.

A day in the life of a lasagne

1 Is the food stored at 4°C?
What is its sell by date?
How long is the food carried around the supermarket getting warm?

2 How long will the lasagne be stored above 4°C? Could it be the rest of the day until the consumer gets home from work, or will they put it into a refrigerator at work?

3 How long does it take the consumer to travel home? What temperature does the lasagne reach?

4 Is the lasagne reheated straight away or is it put into the refrigerator until being eaten later in the evening?

5 Are the reheating instructions being followed correctly? Is there a risk of cross contamination from previously spilled food in the microwave?

6 Is the lasagne thoroughly heated through?
Has the person serving the lasagne cleaned their hands and has good personal hygiene?
Is the plate and cutlery used to eat the lasagne clean?

It can be seen from the flow chart that it is very easy for the temperature of a food product to rise above 4°C and therefore risk the possibility of food spoilage or poisoning.

Summary

You should now be able to:

explain why good personal hygiene is essential to stop food poisoning in food production

draw a simple flow chart for making a high risk product and mark on the safety and hygiene control checks

discuss where food poisoning could occur when buying and reheating a chilled food.

8 Additives

8.1 Types of additives

Food manufacturers use additives in the production of food products for lots of different reasons. The main reasons are to produce high quality food products that have an extended shelf life.

Food additives can be classified as natural or artificial. Natural additives include salt, sugar, spices and natural food colourings and have been used in food preparation for many hundreds of years. Artificial additives are made from a range of chemicals that have been specially designed by food scientists. Additives are only used in small quantities to be effective.

■ What are the main roles of additives in food production?

Additives are used to:

- improve **sensory qualities**, e.g. flavour, appearance, colour and smell
- improve structure, e.g. texture, consistency
- add nutritional value
- increase the time food is safe to eat.

The large numbers of additives used in food manufacture has resulted in a growth in the range of convenience food products available to consumers, e.g. cook chill lasagne, frozen cheesecake, salad dressings. These products are usually of a consistently good quality and often need minimal preparation and cooking.

We often classify additives according to the jobs they do in food production:

- preservatives
- colourings
- flavourings
- emulsifiers
- stabilisers
- anti-oxidants
- nutritional enhancers.

The following food packaging labels show you examples of the main roles additives have in foods.

Objectives

The main types of additives and their role in food production.

The advantages and disadvantages of food additives in our diet.

A Salt is a natural additive

links

Visit the Food Standards Agency at www.food.gov.uk.

Key terms

Food additive: a substance added to a food product to improve its quality.

Sensory qualities: the look, smell, taste, feel and sound of food products.

E numbers: the classification system used for food additives.

Activity

1 Look at four different ingredients lists on food packaging and identify the food additives.

Belgian Chocolate Spread
Ingredients
Sugar, Vegetable oil, Skimmed milk powder (4%), Fat reduced cocoa, Whey powder, Emulsifier (soya lecithin), Flavouring.

! Allergy advice: Contains milk and soya

! Produced in a factory which uses nut ingredient

Nutrition
Typical values per 100g: Energy 2427kJ/578kcal, Protein 2... Fat 38.1g (of ...

Ⓥ **Suitable for vegetarians**

Ingredients
Carrot Cake (73%), Cream Cheese Buttercream Filling (14%), Frosted Topping (13%), Sweet Cinnamon Dusting.
Carrot Cake contains: Wheat Flour, Sugar, Vegetable Oil, Carrot (14%), Egg, Egg White, Humectant (Vegetable Glycerine), Raising Agents (Disodium Diphosphate, Sodium Bicarbonate), Glucose Syrup, Preservative (Potassium Sorbate), Cinnamon, Stabiliser (Xanthan Gum).
Cream Cheese Buttercream Filling contains: Sugar, Medium Fat Soft Cheese (25%), Butter (14%), Glucose Syrup, Cornflour, Modified Maize Starch, Salt, Preservative (Potassium Sorbate).
Frosted Topping contains: Margarine, Sugar, Emulsifiers (Mono- and Di-Glycerides of Fatty

B *Food labels showing the different functions of additives*

We ban smoking yet allow our children to be poisoned with food additives

Food additives make children behave badly

The proof food additives ARE as bad as we feared

C *News about additives appears regularly in the headlines*

Are food additives safe to eat?

Manufacturers, as well as the Government's Food Standards Agency, test all additives that may be used in food products to prove that they are safe to eat. All food additives once tested and found to be safe are given a number and a prefix E; these are known as **E numbers**. These foods can then be sold in the European Union. Highly processed foods can contain large numbers of E numbers.

Consumers are starting to become concerned about the long-term effects these chemicals are having on our bodies as well as babies and young children. Recent studies are drawing links between the increased amount of processed foods eaten by the UK population and the increased amounts of childhood hyperactivity, asthma, eczema and food allergies. The food colouring additives, e.g. E102, E110, E129 are those most frequently linked to childhood illnesses.

Articles in the media have recently highlighted increased concerns about the rise in the amount of food additives used by manufacturers and consumers are starting to demand organic and natural foods from manufacturers. These food products are more expensive to produce and buy as the range of food products available is limited.

AQA Examiner's tip

- Understand what the letter E represents before a number on an ingredients list.
- Know the advantages and disadvantages of using additives in food preparation.

Summary

You should now be able to:

name four functions where additives are used in food production

discuss the need for additives in food production against the possible health risks.

Activities

2 Compare and contrast four types of tomato soup:
- dried
- tinned
- home-made
- microwaveable.

Consider the effects food additives have had on these products.

3 Carry out further research on food additives. Then produce a persuasive argument for the reduction of additives in food products and what the possible effects would be to: the manufacturer, retailer and consumer.

D *Tomato soup, but is it dried, tinned, home-made or microwaveable?*

All additives must be harmless to the consumer but they also need to be flavourless, colourless and odourless so that they do not alter the natural sensory characteristics of the food product unless the manufacturer requires them to do so. Manufacturers have to weigh up the benefits of using additives in food production against the changes to the food product. It must also be remembered that additives can be used to disguise poor quality foods.

How do additives help preserve foods?

Preservatives help food stay edible and unspoiled for longer. They stop the growth of bacteria, yeasts and moulds by changing the environment inside the food so the conditions these microorganisms need to grow are not available. This stops food from deteriorating quickly and extends its shelf life. Preservatives are mainly found in meat products that have been processed, e.g. bacon, sausages. They are also used in baked products, dried fruit, fruit and juice.

Why do manufacturers use additives to colour foods?

Colours are added to food to make it more attractive and appealing to consumers. Colour additives come from both synthetic and natural sources, e.g. E162 beetroot red (natural), E171 titanium dioxide (synthetic). When foods are manufactured the processing may discolour the food, e.g. canned peas and so colours are added to restore the food's natural colour. Colours are also added to foods that otherwise would be colourless, e.g. fizzy drinks or foods that have a hint of colour but could do with the colour boosting, e.g. strawberry milk.

Objectives

Understand how additives are used to extend the shelf life of food products.

Understand how additives are used to make food products more attractive.

A *Preservatives stop bacteria from spoiling foods*

Cheesecake mix ingredients:

Ingredients

Crumb Base Mix (55%): (Wheat Flour, Vegetable Oil, Sugar, Invert Sugar Syrup, Salt, Raising Agents: Sodium Hydrogen Carbonate, Ammonium Carbonates). Filling Mix (45%): (Sugar, Whipping Agent: Glucose Syrup, Vegetable Oil, Emulsifier: Acetic Acid Esters of Mono- and Diglycerides of Fatty Acids; Milk Protein*; Maize Starch, Cheese Powder* (8%), Stabilisers: Sodium Alginate, Tetra Sodium Diphosphate, Monocalcium Phosphate, Calcium Lactate; Dextrose, Whey Powder*, Flavourings, Colour: Beta Carotene).
*FROM COWS' MILK

Homemade lemon cheesecake recipe:

110g digestive biscuits

50g butter

50g caster sugar

350g cottage cheese

2 large eggs

rind and juice of 2 lemons

10g powdered gelatin

150ml double cream

B *Which is most natural? Which is most healthy? Which would you prefer to eat?*

Do manufacturers need to add flavourings to foods?

Flavourings and flavour enhancers are added to foods to improve the flavour so the foods appeal to the consumer. Flavouring additives are used to restore the original food flavours when these have been changed during processing. They are also used to add flavours, e.g. vanilla flavouring to an instant dessert. Flavour enhancers work in a slightly different way, e.g. monosodium glutamate (MSG) – they do not have a flavour of their own but intensify the food's natural flavours. Flavour enhancers are used in many Chinese meals and savoury foods.

Emulsifiers and stabilisers

Emulsifiers and stabilisers are essential in the production of lots of processed foods as they ensure the ingredients do not separate into their component parts. Water based foods do not mix easily with oil/fat based foods and need an emulsifier to ensure they are fully mixed and a stabiliser to ensure they stay in a fully mixed state. Lecithin in eggs is a natural emulsifier. Emulsifiers and stabilisers give a consistently smooth creamy texture and are mainly used in the production of mayonnaise, salad cream and ice cream.

Which other additives are used in food production?

- Sweeteners are a group of flavourings that affect the taste of foods and are frequently used in low calorie and reduced sugar products.
- Thickening and gelling agents are used to achieve a range of textures in processed foods.
- **Anti-oxidants** are used to stop fats going rancid in foods and stop fruit and vegetables going brown.

⚲ links

Emulsification is explained more in 1.3 Protein on page 14.

AQA *Examiner's tip*

- Understand the main roles of additives in food production.
- Know which foods are most likely to contain additives.

Key terms

Preservative: a substance that extends the shelf life of a food.

Emulsifier: a substance that stops oil and water from separating.

Anti-oxidant: a substance that stops fat in food going rancid.

Activities

1. List the types of additives that could be found in the following products:
   a   corned beef
   b   salad cream
   c   Pot Noodle
   d   salt and vinegar crisps.

2. Compare and contrast the texture, colour and flavour of a home-made and shop bought cheesecake.

Summary

You should now be able to:

explain how preservatives work

explain how emulsifiers and stabilisers ensure a good texture for food products.

9.1 Methods

After food products have been developed successfully in the test kitchen they go into large-scale production. **Scaling up** of quantities, sourcing of materials and costings will have been considered to ensure the product will be **commercially viable**. A manufacturing specification will have been written. This gives the precise details that the food manufacturer will need in order to produce an exact replica of the final prototype in the factory. It includes information such as: the types and amounts of ingredients, depth of rolling out, cooking temperatures and times, all of which ensure a high quality product. Designated tolerances for certain measurements will also be given, e.g. +/- 0.5mm for the depth of pastry. The food manufacturer produces a production plan to meet the needs of the specification.

A production plan involves a sequence of activities that result in identical high quality finished products. The plan considers all the ingredients and materials that are needed to produce the product and all the manufacturing processes and activities that make ingredients into a final product. The plan will have checks built into the manufacturing systems that ensure the quality and safety of the final products.

The type of product being made and the quantities required determines the type of production method used to make it on a large scale.

There are four main commercial methods of food production.

One-off production

This is when one food product is made for the specific needs of a consumer, e.g. wedding or celebration cake. It requires specialist skills from experienced workers.

Batch production

This is a small scale production system making a specific number of the same food product, e.g. bread rolls and pasties in a bakery, sausages in a butcher's. Only a small number of people are involved in the food production and they each carry out a specific task, e.g. kneading the bread or shaping the rolls.

Mass production

This is used when large numbers of the same food product are required, e.g. white sliced loaf or breakfast cereals. The manufacturing process is split into single tasks, which are part of a production line linked together by a conveyor belt. All or some of the production line tasks are automated and are carried out using specialist large-scale

Understand the different commercial methods of food production.

Understand the need for production plans that ensure a high quality consistent product every time.

A *One-off production*

B *Food products produced by batch production*

equipment. Computers aid the manufacture of the food products by controlling some of the tasks. Fewer food workers are required in this type of production.

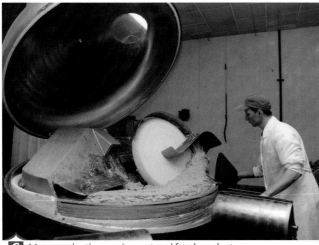

Continuous flow production

This works on the same principle as mass production but always produces the same product all the time usually 24 hours a day, seven days a week. It is used in the production of popular food products, e.g. baked beans, crisps or biscuits. The production line is only used to manufacture one food product and involves expensive investment in specialised machinery. This is an automated process and is an example where computers control the processes.

Activities

1   Draw up a chart to show the similarities and differences between the four main production methods.

2   Which method of production is normally used for each of these products:
   a   salt and vinegar crisps
   b   sausage rolls
   c   Twix biscuits
   d   novelty pizza.

3   In a team of four carry out the batch production of four identical pizzas with each member of the team carrying out one specific task. How similar were your pizzas?

Summary

You should now be able to:

explain the difference between batch and mass production

explain why production plans are important when producing food products.

AQA Examiner's tip

Consider the advantages and disadvantages of different production methods.

Remember

You will not be tested on production methods as part of the controlled assessment.

Key terms

Scaling up: multiplying up proportionally.

Commercially viable: make a profit when it is sold.

Computers are an essential part of the mass production of food products and have a variety of uses from weighing ingredients to **monitoring** the moisture content of food products. Most computer use is in the manufacture of products but they are also used in the designing of new food products.

Computers are used instead of people to carry out lots of aspects of food manufacture because they:

- speed up the process
- reduce error
- **standardise** production
- can be programmed to measure a range of variables that humans can not, e.g. moisture content, consistency
- can work continuously if required to do so
- deal with lots of information efficiently and immediately
- avoid people handling food therefore less risk of contamination.

What is CAD?

CAD is Computer Aided Design.

CAD is the use of computer design programs to produce images or drawings that can be used in the development of food products. The programs help with:

- product profiles
- nutritional modelling of products to meet design criteria
- product modelling of products to meet **aesthetic** and sensory criteria
- production of safety and hazard charts
- mathematical calculations for scaling up
- nets, graphical images, calculations of numbers and charts for the packaging of products.

What is CAM?

CAM is Computer Aided Manufacture.

CAM is the use of computer programs to control and monitor processes during the mass production of food products. The automated processes of mass production are generally known as CAM. The computer has three main jobs:

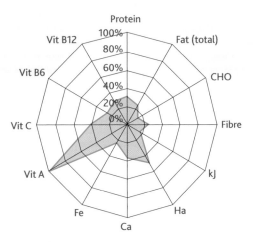

Average intake
Female/Male age 15-50

FOOD for a PC ANZ6.1										
Dairy	Fats	Meat	Fish	Veg	Misc	Fruit	Cereal	Nuts	Drinks	User

	FAT g	CHOg	FIBREg	kJ	SAL Tmg
Consumed as main meal	20	13	5	1609	707
or Appetiser/Dessert	20	13	5	1609	707
or Snack	20	13	5	1609	707

DATA are reproduced by permission of FSANZ and HMSO Version Wustralia and NZ

Producer's Trade name	DIXIE CUISINE
Product	VEGETABLE LASAGNE
Number of portions	4

NEW PRODUCT

Information: METHOD Prepare vegetables and lightly fry. Prepare cheese sauce using all in one method. Assemble in oblong dish in layers. Cover with foil then cook in moderate oven, 40 minutes.

Dairy	Protein g	Fat (tot) g	Fat (sat) g	CHO g	Sugar g	Starch g	Fibre g	k Cal	kJ	Na mg	Ca mg	Fe mg	Vit A µg	Vit C µg	Vit D mg	Vit E mg	Vit B6 mg	Vit B12 µg
100 Lasagne (boiled)	3	3	0	22	1	22	1	100	24	1	6	1	0	0	0	0	0	0
185 Chili Beans	9	1	0	23	5	17	7	130	703	992	113	3	111	0	0	1	0	0
400 Canned Tomatoes	4	0	0	12	11	1	3	64	276	156	48	2	880	48	0	5	0	0
75 Onions (fried)	2	8	1	11	8	0	2	123	513	3	35	1	30	2	0	0	0	0
100 Mushrooms (fried)	2	16	1	0	0	0	3	157	645	4	8	1	0	1	0	0	0	0
120 Zucchiri (fried)	3	6	0	3	3	0	1	76	318	1	46	2	600	18	0	1	0	0
10 Soy Sauce	1	0	0	1	0	0	0	6	27	572	2	0	0	0	0	0	0	0
500 Skimmed Milk	17	1	0	25	25	0	0	165	700	270	600	0	5	5	0	0	0	2
50 Low Fat Spread	3	20	4	0	0	0	0	195	803	325	20	0	953	5	4	3.2	0	0
50 Plain White Flour	5	1	0	40	1	39	2	175	747	1	75	1	0	0	0	0	0	0
75 Cheddar Cheese	19	26	16	0	0	0	0	309	1281	503	540	0	413	0	0	0	0	0.8

A *Examples of CAD in practice*

- to program the machinery to carry out specific tasks, e.g. weighing exact amounts of all the ingredients into a large scale mixer to make pastry
- to prevent food spoilage hazards occurring during the production run, e.g. monitor the temperature of the food during the cooking of a chicken curry
- to monitor the quality of the food product at various stages, e.g. the depth of icing put on the top of a Bakewell tart
- as computers continuously monitor the production run, so any feedback from the computer monitoring that may affect the quality and safety of the final product results in instant changes being made to the production run.

B *Using computer control and monitoring*

How is computer control and monitoring used in food processing?

- Weighing of ingredients to exact amounts.
- Mixing of ingredients for an exact time at an exact speed.
- Chopping of ingredients to an exact size.
- Shaping of a product to an exact shape.
- **Assembling** products.
- Coating products (known as enrobing).
- Cooking of a food product for an exact length of time at an exact temperature.

How can computers monitor food safely?

- Sensor detectors for finding metal or other foreign bodies.
- Microbial level detectors.
- Temperature displays on refrigerators and freezers.
- Critical temperature probes for hot foods.

⚭ **links**

See Controlled Assessment page 146.

> **Key terms**
>
> **Monitoring**: keep constant watch.
>
> **Standardise**: make everything the same.
>
> **Aesthetic**: attractive.
>
> **Assembling**: putting component parts together.

> AQA **Examiner's tip**
>
> - Understand the difference between CAD and CAM.
> - Understand why using computers results in a consistent product each time.

> **Activities**
>
> 1. List the processes you think CAM would be used to carry out in the mass production of pizzas.
>
> 2. Carry out the nutritional modelling of a pizza to meet the needs of a target group.

> **Summary**
>
> You should now be able to:
>
> describe how CAD could be used in the production of a food product
>
> list the differences between CAD and CAM
>
> give examples of how CAM is used in mass production of food.

9.3 Quality control

Consumers like to know that each time they buy a product it is identical to one purchased previously, e.g. a tin of tomato soup, packet of crisps, frozen vegetable risotto. To ensure that this happens, food manufacturers put in place quality control checks throughout the making of the product. These checks cover a wide range of variables that can happen when a food product is being produced, e.g. undercooking, incorrect shape, too sweet. Some of the checks have to be within a **tolerance**, e.g. +/− 0.5mm for the depth of pastry, but some need to be exact, e.g. length of whisking time for meringue otherwise the product will be of poor quality.

A food handler carries out **visual** checks and computers carry out more scientific checks. Sensors are attached to a computer, which link to the actual food or the food production machinery or the surrounding atmosphere. This information is fed back into the computer and changes, no matter how small, are made to the production processes. An alarm will sound if there are problems with the production.

What sort of monitoring is carried out by computer sensors?

- The temperature of the food at all stages of the production run.
- The pH levels.
- The moisture content.
- The thickness of the dough.
- The colour of the food.
- The weight of ingredients and the food product.
- The rate of flow for adding liquids or fillings to products.
- The timing of cooking and cooling.

In most food production factories **sensory evaluations** are also carried out on every production run at least once a day. These look at the sensory characteristics of the product and whether they meet the specification, e.g. taste testing.

Objectives

Understand the importance in commercial food production of making high quality products every time.

Understand the range of quality control checks used in food production.

A *Computers monitoring food production*

B *Sensory evaluation during food production*

⃝⃝ links

See Chapter 13 Sensory testing.

Activities

1 List the quality control checks you think should be in place when making samosas.

2 In a team of four write a production plan with quality control checks for making fairy cakes in the food technology room. Each of the team follows the plan and see if your final products are identical. If not why do you think this happened?

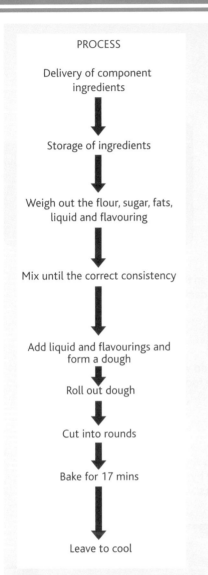

PROCESS	QUALITY CHECKS
Delivery of component ingredients	are the ingredients of good quality and well within their best before/use by date? Is the packaging damaged?
Storage of ingredients	are the storage areas clean, dry and at the correct temperature?
Weigh out the flour, sugar, fats, liquid and flavouring	are the ingredients measured accurately? Have digital scales been used to check tolerances?
Mix until the correct consistency	is the mixer at the correct speed, using the correct attachment, for the correct length of time?
Add liquid and flavourings and form a dough	is the consistency correct?
Roll out dough	is it the correct depth?
Cut into rounds	are these identical shapes?
Bake for 17 mins	is the oven at the correct temperature, for the correct length of cooking time? are the biscuits the correct colour?
Leave to cool	are the biscuits of the correct shortness?

C *Simple production flow chart for making a biscuit*

AQA Examiner's tip

- Understand why quality control is essential in food production.
- Be able to give examples of types of quality control used for a range of products.

Remember

As part of the controlled assessment you will need to produce a flow diagram to show how the final product will be made in the test kitchen. You will need to show where you have considered quality control.

Key terms

Tolerance: the amount of difference allowed when making.

Visual: looking at something.

Sensory evaluation: using the range of senses to asses a food product, e.g. appearance, smell, taste.

Summary

You should now be able to:

list six quality control checks frequently carried out in the mass production of foods

explain why it is important that food manufacturers always produce identical high quality products.

10.1 Packaging information

Objectives

Understand what information is present on packaging.

Understand what information is legally required to be on packaging.

Most food outlets are self service and because of this it is important foods are packaged.

Why we package food

Labelling and the law

It is a legal requirement to inform consumers about the food products they are buying. The Food Labelling Regulations of 1996 state the type of information that must be displayed on a food product label. There are eight requirements.

The food product name

- This must include the type, e.g. curry.
- If it has been processed, e.g. smoked bacon.
- The pictures must not **mislead** the consumer.

List of ingredients

- All ingredients must be listed.
- Ingredients are listed in **descending** order of weight starting with the largest.
- Food additives must be included and the category name written before each additive, e.g. anti-oxidant.
- Unwrapped foods, e.g. cookies need to have information regarding allergies and additives at the point of sale.

Weight or volume

- Foods if pre packed must display the **net weight** or volume.
- If not the quantity or number can be used.
- The symbol of an **e** covers the average minimum amount in a packet.
- Weights under 5g do not need to be stated.

Instructions for use

- These must state how to prepare and cook the food.
- Cooking times and temperatures must be given.

Protection from bacterial damage.

Protection from physical damage.

Contains the food product.

Preserves the product to extend its shell life.

"Why do we package food?"

Informs the consumer about all aspects of the product from storage instructions to weight.

Attracts consumers to buy the product.

Stops **tampering** by other consumers.

Eases storage.

Eases transportation.

A *Why do we package food?*

B *An example of food packaging*

- Instructions often include microwave and conventional cooking.

Storage instructions

- These must state how best to store the food to prevent food spoilage.
- Temperature guidelines must also be given.

Use by date/Best before date

- These are used to tell the consumer how long a food should be kept to ensure it is of good quality when it is eaten so that food poisoning does not occur.
- Use by date is used for high risk foods and this is the date the food must be eaten by.
- Best before date is used for lower risk foods and states the date the food will start to deteriorate in terms of taste, flavour, colour and texture.

Name and address of the manufacturer

This is a contact for consumers who may wish to make a complaint about the food product; it can be either the manufacturer or the supplier's name and address, e.g. Kelloggs or Morrisons.

Place of origin

This shows the place the food has come from and is becoming more important as consumers consider the environmental issues of food travelling around the world before we eat it.

Allergic information

This shows the ingredients in the food that consumers may be allergic to, e.g. nuts, eggs, gluten.

Other labelling information often present on food packages

Nutritional information

This information is given in a variety of ways; originally the amount of key nutrients per 100g as well as the amount per average serving.

The Foods Standards Agency has devised a traffic light system to help consumers see the levels of fat, sugar and salt contained in ready-made foods more easily. Red means high, green means low and amber means moderate. The quantity of the fat, sugar and salt is also given per serving.

A lot of food manufacturers have devised their own front of package labelling system which shows calories, sugars, fats and salt in each serving as a percentage of the guideline daily allowance. Consumers cannot see quickly from this system if the food is particularly high in any of the stated nutrients.

Special information

This gives information to attract the consumer to the product and includes:

- foods that may be suitable for a specific diet, e.g. vegetarian
- foods that can be cooked or stored in a certain way, e.g. suitable for freezing, suitable for reheating in a microwave oven
- packaging that can be recycled or that has been made from recycled products.

⬭⬭ links

See 2.3 Nutritional labelling.

Key terms

Tampering: to interfere with.

Mislead: not telling the truth.

Descending: from the largest to the smallest.

Net weight: not including packaging.

Activities

1 Design your own packaging net for a snack product and add the legally required labelling to it.

2 Look at examples of the different types of nutritional labelling on foods and decide which you feel is most helpful to a consumer and explain why.

Summary

You should now be able to:

design and annotate your own packaging

list five reasons why we package food.

kerboodle!

10.2 Packaging materials

Food is packaged for many reasons and consumers are becoming more aware and concerned about the effect food packaging is having on the environment. The amount of waste that food packaging produces every year is huge and a lot of it is non-recyclable. Food manufacturers are aware of this and are developing more environmentally-friendly packaging.

Packaging materials are becoming more and more specialised as technology advances but must not:

■ be dangerous to human health
■ cause food to **deteriorate**
■ cause unacceptable changes in the quality of the product.

The four main groups of packaging materials are:

■ metals
■ glass
■ card and paperboard
■ plastics.

Metals

A *Foods packaged in metals*

The two types of metals used are tinplate and aluminium.

Advantages:

■ strong
■ used in different thickness, e.g. foil and tin cans
■ can be moulded into different shapes
■ lightweight
■ **impermeable** to contamination
■ can be recycled
■ preserves food
■ can be heat treated.

Disadvantages:

■ can react with some foods so some cans need non-metallic linings.

Objectives

Understand the range of packaging materials available.

Understand the functions and properties of packaging materials.

Key terms

Deteriorate: starting to decay and losing freshness.
Impermeable: cannot penetrate.
Biodegradable: broken down totally by bacteria.

⊂⊃ links

See Chapter 7 Food safety and hygiene.

Glass

Glass bottles and jars are used to package some foods.

Advantages:

- can be moulded into various shapes
- rigid
- transparent so the product can be seen
- can be recycled
- resistant to high temperatures when the foods are being added
- impermeable to contamination.

Disadvantages:

- fragile and easily broken
- heavy.

B *Foods packaged in glass*

Card and paperboard

Card and paperboard are used for packaging dry products unless they are coated with plastic.

Advantages:

- can be printed on easily
- can be made in various thicknesses
- can be moulded and folded into different shapes
- can be laminated or coated
- lightweight
- cheap
- can be recycled
- **biodegradable.**

Disadvantages:

- can be squashed and the contents damaged
- not water resistant.

C *Foods packaged in card and paperboard*

Plastics

There are a large range of plastics used for packaging that each have their own qualities.

Advantages:

- can be moulded into a range of shapes
- lightweight
- impermeable to contamination
- cheap
- can be easily printed on
- water resistant
- can be rigid or flexible.

D *Foods packaged in plastics*

kerboodle!

Thermoplastics

These become soft when heated and can then be moulded in to the required shape.

- PP is polypropylene used for plastic wrap and containers for ready prepared meals.
- PS is expanded polystyrene used for trays and containers. The polystyrene is expanded and shaped. It is a poor conductor of heat and often used for takeaway products.
- PET is polyester used as film to cover the top of moulded plastic trays that reheat food in the microwave as well as fizzy drink bottles.

Which packaging is best for which products?

It is important that the food manufacturer considers the properties of the food product they want to package and also what the functions of the packaging are required to be.

- Is the food a liquid or a solid?
- Is the food fragile or robust?
- Is the food light or heavy?
- Is the food high, medium or low risk?
- At what temperature will the food be stored?
- What is the shelf life of the food?
- Does the packaging need to extend the shelf life of the food?
- Does the packaging need to be printed on?
- Does the packaging need to be rigid or flexible?
- Does the packaging need to be reheatable?
- Does the packaging need to be cheap?
- Will the packaging react to the foods it is containing?

E *What are the requirements of packaging?*

Specialist packaging of foods

Vacuum packaging is a way of preserving food that has been used for many years. The food is placed in plastic packaging and the air around the food is then sucked out and the plastic bag sealed. The food is now in **anaerobic** conditions. Once the food is opened the food needs to be stored in normal storage conditions. Bacon, fish and coffee are most commonly packaged this way.

Modified atmospheric packaging (MAP) is used to extend the shelf life of food. The process involves placing the food in its plastic packaging and replacing the air with a mixture of oxygen, nitrogen and carbon dioxide. The plastic bag or lid is **hermetically** sealed and then stored in chilled conditions. The advantages of this method of preservation are:

- a wide range of foods are stored this way, e.g. meat, fish, salads, fruit, fresh pasta
- the colour of the food stays the same
- an increase in shelf life of high risk foods, e.g. meat for up to seven days if chilled.

Aseptic packaging is used to preserve foods without using preservatives or chilling. The food is sterilised outside the packaging by heating for 3 to 15 seconds. The sterilised food is then placed in air tight sterilised packaging in a hygienic environment. No air gets to the product once it is sterilised. It is used mainly for liquids and as an alternative to canning and the foods keep for up to six months.

Key terms

Anaerobic: not needing oxygen.
Hermetically: airtight.

Activities

1. Look at the different ways milk is packaged and discuss the advantages and disadvantages of each method. Remember to carry out sensory evaluation as well.

2. Make a main course product using pre-packaged foods.

3. Carry out a supermarket survey to see how many different types of packaging you can see and the types of foods packaged in each way.

Summary

You should now be able to:

match a food to its best packaging material

understand the principles underlying recent methods of food packaging.

A lot of media attention is focused on the effect we as consumers, food manufacturers and retailers are having on the environment now and are going to have over the next fifty years. The main concerns regarding the packaging used for foods are:

- It can not always be **recycled**.
- It is not always biodegradable, i.e. rots down to become compost.
- It uses up the world's natural resources of oil, metals and trees.
- During the production of packaging air, land and water pollution occurs.
- Transportation of the packaging around the world causes air pollution.
- 90% of rubbish is put in **landfill sites**.

What can manufacturers do about this?

- Use recyclable and biodegradable packaging wherever possible.
- Ensure consumers know which packaging is recyclable by using recycling logos on their packaging.
- Use less packaging.
- Stop pointless packaging, e.g. shrink-wrapped cucumbers.
- Packaging could be made thinner.

What can we do as consumers about this?

- Choose products that come in recyclable or biodegradable packaging to encourage manufacturers to package food in this way.
- Ensure we recycle as much of our food packaging as possible; up to 70% could be recycled but consumers only recycle 33%.
- Buy a single larger size rather than individual portions, e.g. yogurt.
- Use our own shopping bags instead of plastic carrier bags.

Scientists and technologists are developing biodegradable food packages based on starch based waste products. These include corn starch and potato starch.

What are genetically modified foods?

These are foods that scientists have altered the **genes** within the food to give it other characteristics. Combining or adding genes from one food to another achieves this. The characteristics achieved can:

- make crops resistant to disease
- increase the nutritional quality of a food
- increase the quantity of the food grown from the same amount of land and seeds.

GM foods were thought to be the answer to lots of the malnutrition and starvation in the world. However, some scientists feel that genetic engineering could lead to side effects in other foods grown near GM foods. More research is required as to the long term effects on health before consumers will accept GM foods are safe to eat. GM ingredients have to be specified on food labels.

B *GM foods*

What are nanofoods?

All products are made from atoms. These are between 1 and 100 nanometres (nm) in size – that is one millionth of a millimetre. The properties of food products depend upon how these atoms are arranged. If we rearrange these atoms we can change the food product completely, e.g. if we rearrange the atoms in soil, water and air we can make potatoes. This technology is very new, still at an experimental stage and could have a massive impact on food production. It could allow us to make packaging materials that are cleaner, stronger and lighter, to reduce the amount of fat, salt and sugar in foods but still keep the flavour and mouth feel. As with GM foods there are great concerns about the side effects on our health from processing foods.

What are organic foods?

These are foods that have been grown without the use of chemicals, fertilisers or pesticides. Organic foods contain no artificial additives. Consumers who are concerned about the long-term effect of additives and chemicals in our food on our health want to eat organic foods. Food manufacturers are developing lots more organic foods as demand increases, however they are expensive.

What are Fairtrade foods?

Fairtrade foods make sure that the workers who produce the foods get a fair price for their products and have a reasonable standard of living in the developing world. Fairtrade is about guaranteed fair prices for the farmers, farm workers and their families, better working conditions and local **sustainability**. Companies who buy the farmers' products must pay the market price. Fairtrade products are becoming very popular in the UK with the amount of products sold doubling in value every two years at present.

What are Farm assured foods?

Farm assured foods have been produced to meet specific standards for homegrown foods. The food must be farmed and packed in the UK and meet strict hygiene, safety and welfare standards at all stages. These standards are stricter than the legal minimum set by experts from the Government and food industry.

What are food miles?

As concerns about our environment grow it is recognised that the production of CO_2 is a main contributor to global warming. Transport produces a lot of these emissions. The distance foods travel from where they are grown to where they are bought by the consumer is known as food miles. Some foods travel tens of thousands of miles before they arrive in the supermarket, e.g. bananas. As consumers demand all year round availability of foods, supermarkets source their produce from around the world. Another reason is that foods produced in some countries are a lot cheaper than in the UK. One way to cut down on food miles and therefore reduce pollution is to eat foods produced locally and to eat foods that are in season.

C Organic foods

D Fairtrade logo

E Farm assured logo

Activities

1   Write a newspaper article about how our food choices are affecting the environment.

2   If possible visit your local supermarket and carry out a survey of food products to see how many of the logos described in this chapter you can see, also look at the fresh produce and see how far it has travelled.

Summary

You should now be able to:

explain the impact of two new initiatives on food quality and sustaining the planet

explain how more efficient and effective use of packing can contribute to protecting the world's resources.

kerboodle!

AQA Examiner's tip Questions will require you to show you know and understand that food products are developed and made safe to eat by combining different ingredients and by using a range of different processing methods and equipment.

1   (a)   (i)   What is meant by a 'control check'?   *(3 marks)*

           (ii)   Explain why some control checks are done by computers.   *(4 marks)*

           (iii) Describe different control checks that are used when making bread rolls.

Preparation stage	Control checks used
Choosing raw ingredients	
Preparing ingredients	
Cooking a food product	

(6 marks)

     (b)   Control checks are made on different batches of salmon and prawn filo parcels.
Problems found during the checks are listed below.

           (i)   Give one cause of each problem.   *(5 marks)*

           (ii)   Explain how the problem may be prevented.   *(5 marks)*

Problem	i) Cause	ii) How to prevent this problem
Filo pastry is dry and breaks up when handled		
The filling leaks out of the parcel during cooking		
Pieces of shell are found in the filling		
Creamy sauce is thin and runny		
The final product is pale and lacks colour		

(10 marks)

2   A local takeaway outlet is developing a new range of vegetarian meals.

     (a)   Sketch two **different** design ideas for a takeaway food product.

           * Do **not** draw any packaging.

           Annotate your sketches to show how your design ideas meet the following design criteria:

           Criterion 1: offers a healthy option

           Criterion 2: provides a variety of textures

           Criterion 3: provides a hot and spicy flavour

           Criterion 4: suitable for vegetarians   *(12 marks)*

(b)   Choose on of your design ideas.
        Design idea 1 ☐   Design idea 2 ☐

   (i)   Using flowcharts, notes and/or diagrams produce a plan for making your
         chosen design idea in the test kitchen.   *(10 marks)*

   (ii) Describe the personal health and safety rules to be followed by food
         workers following your production plan.   *(8 marks)*

(c)   Explain why the following may be used when cooking and serving your
        food product:

   (i)   colour coded food preparation equipment

   (ii) a food probe.   *(6 marks)*

3   (a)   Complete the table below to show key temperatures used by manufacturers
                  to ensure food safety. Use these temperatures:

          72°C   5–63°C   72°C   37°C   −18°C

Temperature	
	bacteria are dormant and not able to reproduce
	bacteria are 'sleeping' and reproduce very very slowly
	bacteria reproduce most actively; this is known as the danger zone
	optimum temperature for bacteria to reproduce
	bacteria start to be destroyed and are not able to reproduce

                                                                          (5 marks)

 AQA
Examiner's tip These questions are based on general food products but similar questions may be asked on
any chosen food product. You may like to practise similar questions for the food product of
your choice.

AQA specimen question

Design and market influences

This section is about the stages in the development of a new food product. The development follows a process called the 'design process'. This process will be familiar to you because you will have used it in Key Stage 3 when you designed and made food products.

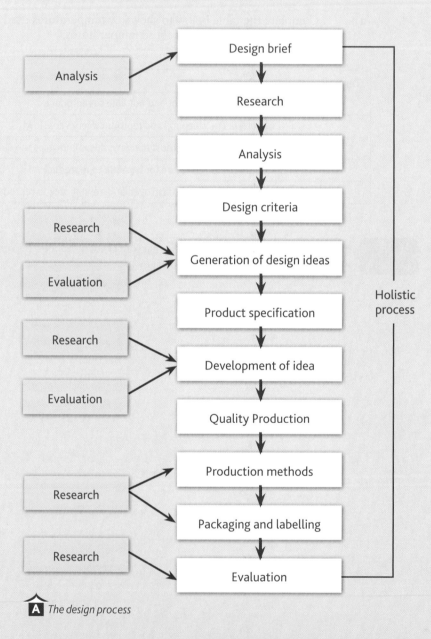

A *The design process*

New products do not appear on the shelves of supermarkets or on the menus of restaurants by chance. A lot of time, money and resources are invested into the development of new food products.

What will you study in this section?

Chapters 11–16 within this section address aspects of the design process. Explanations about the purpose of each stage with examples are provided, together with suggestions and explanations about how to carry out some of the different aspects of designing and making. The chapters mainly focus on making a product in a test kitchen but also included are aspects of larger scale production and manufacturing which you will need for the written examination. The section also gives suggestions about the 'making' part of the design process and links to the sections on Materials and Components and Processes and Manufacture.

How will this unit help me?

In order to carry out your 'Designing and Making Practice' and to be able to answer questions in a written examination you need to have gained knowledge, understanding and skills. This unit provides you with understanding and knowledge about the design process which is key to the Controlled Assessment but also will enable you to apply that knowledge and understanding when you have to answer examination questions. The skills you will have learned through explanations and examples in this section relate to both designing and making. This should enable you to apply these to the process of developing your own product and also to questions in the written examination.

kerboodle!

11.1 Analysing a design task

■ What is a design task?

A **design task** sets out a situation or customer need for a food product. You will be using different design briefs during the course and also for your final piece of coursework. Some design briefs are very short and focused and give a very clear message about the food product which you are to design and make. The task below illustrates this:

'Design and make a savoury main meal product which incorporates a sauce.'

You know from the beginning the type of food product you are being asked to design and make.

Other tasks are very wide and general as the one below shows:

'Design and make a product which is suitable for a special diet'

You need to re-word the task by choosing s specific diet as the examples below show:

- Design and make a main meal product for a vegetarian.
- Design and make a dessert for someone who is 'lactose intolerant'.

How to choose a design task

Think about the following:

- Are you interested in the task?
- Can you find relevant information?
- What ingredients are you likely to need?
- Do you think the task will enable you to make a variety of different ideas?

■ Analysing design tasks

Once you have a design brief you need to start thinking about what the task is asking you to do. Some of this will come from key words; other aspects will come from your own thoughts about the task. This is called **analysing the design task**. Analysis should be quite short and concise and include specific rather than general areas to consider.

How to analyse a design task

There are often clues in a design task which can be the starting point for your analysis. The task will also trigger your own thoughts and suggestions.

The example below shows one way of analysing a design task and presenting your information:

A *Choosing the correct design task*

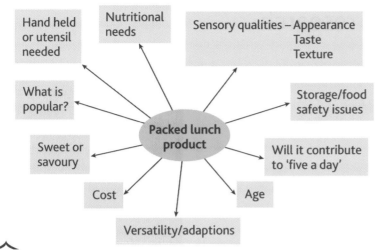

B *Design and make a packed lunch product for a child*

There are other ways to present your analysis, e.g. a table, a bullet pointed list of points or a short explanatory paragraph.

Analysis needs to be clear as this informs the amount and type of research you need to carry out.

Examples of other design tasks

Design and make:

- a main course product for a low fat diet
- a product which makes a good contribution to 'five a day'
- a high quality cake to be sold in a local coffee shop
- a product to be included in the supermarket's 'Around the World' range.

Activities

1   Design and make a complete meal for one which can be sold from the freezer.

2   Design and make a product to be sold in a vegetarian restaurant.

3   Highlight key words and produce a spider diagram to explain what you are being asked to do and what your next steps will be.

links

See Controlled Assessment Section page 138.

Key terms

Design task: a statement which provides the situation for your designing and making.

Analysis of task: splitting up the design task to identify key points.

AQA Examiner's tip

- When analysing your design task provide short, clear and relevant points to highlight key aspects, with explanations if appropriate.

Remember

Design tasks are your starting point. They give you opportunity to be creative.

Summary

You should now be able to:

choose, adapt and use a design task

analyse a design task.

Carrying out research

What is research?

Research means finding out information to help you respond to a design task. Research must always be relevant to the design task and must help you move forward with the design and make process.

Research methods

There are many research methods which you can choose to use. The following list gives you some ideas:

- Product surveys and comparisons give information about what is currently available including size, cost, etc. A short general summary of your findings is the only requirement and not a two page list of every shop and product you have found.

- **Product analysis** sometimes called product appraisal – it gives information about ingredients, sizes, cooking times, storage, packaging and labelling and sensory characteristics.

- Ingredient information and investigations – this enables comparisons to be made between different ingredients and their functions.

- Nutritional information and analysis – this is very important if you are considering a specific target group or diet.

- Recipe analysis and trials – this provides information about making methods, ratios and proportions and in addition gives opportunity for some 'making'.

Questionnaires – these need to have a clear aim related to the task and be well thought out. Only ask questions which give answers you can use and ask a reasonable number of people – asking three people will not give useful information.

A questionnaire should only be carried out if it is relevant to the design task and not because it is a good method of research. Often students do not ask questions which produce any relevant information.

- The following shows the types of questions which could be included in a questionnaire together with the reasons for asking these questions.

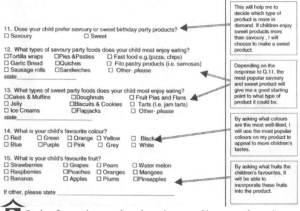

B *Style of questions and explanations used in a questionnaire*

Objectives

Understand methods which can be used for research.

Understand how to use the methods.

Understand why research is needed.

Understand where information can be found.

Understand when to carry out research.

A *Products in a supermarket*

∞ links

See 17.1 Research, analysis and criteria.

There are advantages and disadvantages of using each of these methods but you must remember that you will be given credit for using relevant research methods to produce concise research information. This must be related to the design brief and must have an impact on what you are going to do. Using all the above methods is unnecessary and would result in:

- too much information
- much being of little use to your designing and making
- being very time consuming to carry out and write up
- taking up far too much space in your design folder.

C *Using the computer to collate results*

D *Making a product*

Why research is needed

You will be working to a design task which requires you to have some knowledge about a particular type of food product. You will already know quite a lot about designing and making food products from your studies and therefore you should use this information. This is called using **prior knowledge**.

Design task – bread product

> 66 *I visit the supermarket and local bakery regularly and know the vast range of types of bread and bread products that are on sale. We are now influenced by other cultures and other countries in the types of bread products we buy. I am also aware of the contribution that bread can make to our diets both nutritionally and for adding variety and enjoyment. Most shops sell British, French, Italian, Indian and other bread products so finding a gap in the market will be difficult.*
>
> *I have made bread many times before and already know the types of flour, other ingredients and methods needed to make a good quality loaf of bread. I also know ingredients that are used traditionally in breads from other countries. In order to find out more about what the public like, dislike or need in a bread product I will need to carry out a survey. I can base my questions on what I already know.* 99

Key terms

Product analysis: examining and disassembling an existing food product.

Questionnaire: a series of questions asked to a range of people. The results can be used to inform ideas.

Prior knowledge: knowledge you already have without the need to find out.

Target group: the specific group of people at which you are aiming your product.

You may need additional information using the methods described earlier. You therefore have to decide what additional information you need to find out, where you will find that information and how you will use the information. The following are some sources of information:

- Communicating with people including **target groups.**
- Looking at products in shops and food outlets.
- Looking in books and magazines and on the internet.
- Examining existing products.
- Making products from existing recipes.

When do you do research?

Students often make the mistake of doing lots of research at the beginning of their project whether they need it or not! Research can take place at many different points during the designing and making of a food product.

Initial research

This refers to research which is done at the beginning of your project. It should generate information which will allow you to produce a design specification and design ideas.

Further research

This could occur at any point in the design process.

- Design ideas stage – recipes and methods.
- Development stage – product analysis, functions of ingredients, sensory testing methods, nutritional analysis.
- Manufacturing stage – packaging materials, labelling and standard components.

E *Recording the results from taste tests*

The table below is a good starting point for you to use in your research.

F

What do I want to find out?	What do I already know?	When do I need to find it out?	Where do I find the information?	What methods will I use?

How do I present my research findings?

Research should be presented on a limited number of sides of paper therefore you need to use the space well on each page. You have to plan how you are going to present your findings. The following are some examples of methods you might use:

- short paragraphs of text
- bullet pointed lists
- tables
- graphs
- charts
- labelled diagrams.

Activities

1. Working in pairs, design a questionnaire which will provide you with information about 'celebration products'.

2. Consider the following: 'Design and make a product suitable for a buffet party.'

Using the table above produce a list of what relevant research will be required. How you will carry it out and at what stage in the design process will it occur?

Summary

You should now be able to:

write good questions for a survey

collate and present results

choose relevant research methods and carry out the research

present information concisely.

AQA Examiner's tip

- Always use relevant questions not general ones in your questionnaire.
- Try to be very concise in the way you present results and conclusions in your folder.
- Always show how you will use your conclusions.
- Present a summary only of the information you have collected.
- Do not describe the research methods you have used.
- Carry out research at appropriate points in the process not all at the beginning.

Developing criteria

Analysis of research

When you have completed your research and presented it in your design folder it is important to pick out the most important points of information which will inform your product development. We call this analysing the research.

The following analysis has been produced in relation to a design task requesting a product for a 'party pack of multicultural foods'.

Shop survey

After carrying out a shop survey of party pack products in four supermarkets I found out the following:

Price range: 99p to £2.99.

Most popular products: savoury products to include samosas, spring rolls, prawn toast, mini pizzas, mini poppadums, sausage rolls, vol au vents, mini quiches, bruchetta.

Portion size: hand held individual mouth sized portions.

Nutritional content: high fat mainly saturated, low protein, high unrefined carbohydrate, high salt.

Suitability for vegetarians: many contained animal fat and protein.

Storage: most were sold as cook-chill or frozen.

Target group: no specific target group.

Sensory analysis of existing products

'I tested samosas, mini pizzas, bruchetta and sausage rolls.

The characteristics were:

- Highly seasoned with salt.
- Textures made them difficult to eat with fingers and without a plate.
- Fillings/topping were difficult to keep in the hand.
- Pizza dough was tough.
- Sausage meat lacked flavour.
- Samosa pastry was greasy.
- Most products tasted better after being reheated.'

Magazine and internet research

Looking in magazines and on the internet I found that a greater range of party foods could be included in the pack including sweet products.

Questionnaire

'From the results of my questionnaire I can draw these conclusions:

- Party pack should be for teenagers.
- Majority of people preferred pizza to be included, with spring rolls coming in second place.

This pizza dough is tough!

A *Sensory testing*

Objectives

Understand how to analyse information from research.

Understand how to write design criteria.

Activity

Make a list of some of the general characteristics a main meal product might need to include.

∞ links

See Controlled Assessment page 141.

Key terms

Specification: details which describe the desired characteristics of a product.

Design criteria: a list of general points from which a range of different ideas can be made.

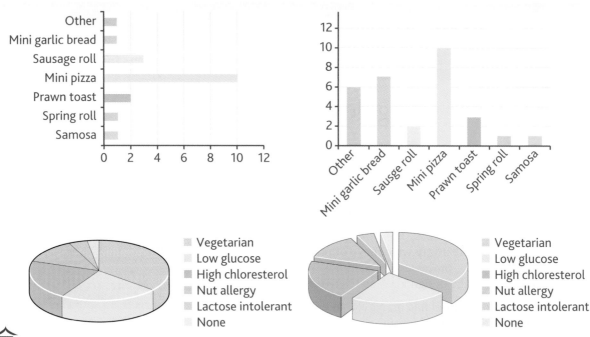

- Majority would use the pack for a birthday party.
- Products in pack need to meet dietary needs particularly vegetarians.'

Writing design criteria

Whenever a new product is being made it has to meet the requirements of a **specification**. After you have carried out and analysed your research you need to produce **design criteria**. This is a list of very general points from which a range of different ideas can be generated. It is very important that you use the information from your initial research and analysis to write your design criteria. Sometimes this list is called a general specification or a design specification.

Using the information from the above analysis of research, the following design criteria can be written:

- easy to eat hand held food product
- sweet or savoury product
- suitable for vegetarians
- costs no more than '£2 for 12'
- targeted at teenagers
- well flavoured
- low in salt
- low in fat
- good texture
- interesting
- can be reheated.

As you can see these are very general points and would enable a range of very different ideas to be generated. There is no mention of particular products, ingredients, methods, flavours, etc. This is why we call it design criteria.

AQA **Examiner's tip**

- Always use information which you have gathered from research to inform design criteria.
- Make sure that your design criteria enable you to generate a range of different ideas which you could make and which would use a range of skills.
- You will need to use design criteria to help you design a given product.

Remember

You need to analyse all the research information you have produced not the research methods.

Summary

You should now be able to:

analyse your research

use findings from your research to write your own design criteria.

12.1 Generating design ideas

Using design criteria to generate design ideas

On the previous page you learned about how to write design criteria and how these could be used to generate a range of quite different design ideas. Design criteria should only include general points.

Design ideas should include a very wide range of products which use many different ingredients, processes and methods. You should also think about how each of the design ideas could be developed.

If for example your design brief had asked you to design and make a dessert then your criteria should be sufficiently general to enable you to generate ideas for: hot and cold, different sizes, flavours, costs, methods of making, etc.

How many design ideas do I need to produce?

There is no rule about how many ideas you need to generate, however present sufficient to show a good range of products using a range of skills, ingredients and opportunities for development.

Communicating design ideas

Remember that you are trying to **communicate** your thoughts to other people. Therefore you must be able to present your ideas in a form which people can recognise. Some of the most common ways of communicating ideas are:

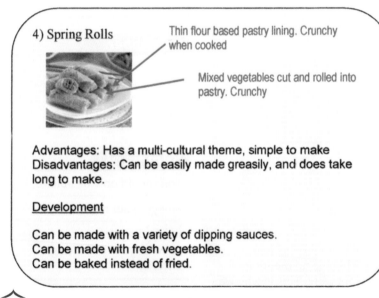

4) Spring Rolls

Thin flour based pastry lining. Crunchy when cooked

Mixed vegetables cut and rolled into pastry. Crunchy

Advantages: Has a multi-cultural theme, simple to make
Disadvantages: Can be easily made greasily, and does take long to make.

Development

Can be made with a variety of dipping sauces.
Can be made with fresh vegetables.
Can be baked instead of fried.

A *Different ways of presenting design ideas*

- sketches or drawings
- written descriptions
- photos or pictures from magazines
- printouts of products from the internet
- photos of products from packaging or advertising.

Annotating design ideas

The most important aspect about presenting your design ideas is the information or thoughts you are having about each one. We call this annotating design ideas.

Annotation should not just describe what you can see but should give additional detail about each idea and most importantly give suggestions about how each idea could be developed.

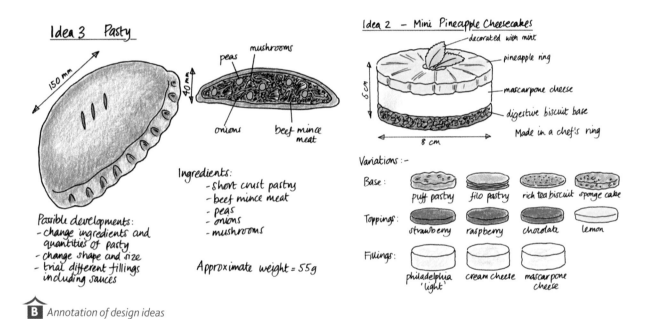

B *Annotation of design ideas*

Activity

Choose a partner and look at the ideas they have generated from their design criteria. Act as a critical friend and produce a feedback sheet explaining how they could improve their work.

Summary

You should now be able to:

use design criteria to produce design ideas

find different ideas

present ideas in different ways and include relevant information about each idea.

AQA Examiner's tip

- Generate ideas which include a range of products, ingredients, processes and skills.
- Ensure each idea is clearly annotated with your own ideas and comments.
- You will need to annotate your design ideas to explain how they fit the criteria given... practice, practice, practice!

kerboodle!

12.2 Choosing design ideas

Which ideas shall I make?

'Making' is a very important part of the design process. Design ideas give you an excellent chance to show your 'making skills'. Therefore you need to think carefully about choosing which ideas to make. Think about products that:

- are different
- use a range of skills, e.g. a crumble is a very simple product using a limited range of skills whereas a lasagne uses a wide range of skills
- can be finished to a high standard, e.g. a fruit tart with a custard filling and a fruit glaze – this not only uses a range of skills but needs accuracy and care in its presentation
- demonstrate quality control, e.g. a batch of samosas need to all be the same in terms of shape, size, colour, etc.
- include different flavours and ingredients
- do not use standard components
- include different preparation and cooking methods
- include **higher-level making skills.**

How do I record in my folder what I have done?

For each idea that you have made you need to be very concise when presenting information. The list below suggests the maximum amount of information you need to present:

- name of product
- list of ingredients used with quantities
- some form of sensory evaluation
- a photograph showing the quality of finish
- nutritional analysis **only** if it is included in the brief or criteria, e.g. low fat, product for a child's lunch, high fibre, low sugar, etc.
- costing if included in the criteria
- **no** plan for making is required.

Objectives

Understand how to decide which ideas to make.

Understand how to use existing recipes and methods.

Understand that making uses a wide range of ingredients and skills.

Understand how to make high quality products and achieve a very good finish.

Understand how to record the making of design ideas.

⃝⃝ links

See 14.1 Evaluating design ideas.

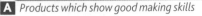
A Products which show good making skills

Pineapple Roulade

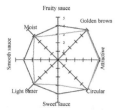

Ingredients:
3 eggs (room temperature)
75g flour
75g caster sugar
125ml whipping cream
10g pineapple chunks

Piped cream with pineapple make it look more attractive as I saw in my research

Dusted with icing sugar

Cream in the centre

No cracks in the roulade, spongy texture

A Radar Chart To Show the Intended Outcome Vs The Actual Outcome of the Pineapple Roulade

DESIGN IDEA 2: Orange and ginger pancakes

Desired	———	
Actual	———	

Ingredients: 125g plain flour, 2 eggs, beaten, 25 cl milk, 1 tbsp caster sugar, 30g butter + 30g for frying, 1 pinch of salt, 2 oranges, 1/2 pot of marmalade, powdered ginger, icing sugar

Method: 1 - Sift the flour and salt into a bowl. Make a well in the middle and add the beaten egg, sugar and 1/3 of the milk. Mix vigorously. Gradually add the rest of the milk.
2 - Melt the butter, and add to the battle. Mix, cover and then leave to stand for 2 hours at room temperature
3 - Peel the oranges, and separate the segments. Place in a bowl, ginger powder and mix. Marinade in the fridge for 30 minutes, to release the flavours.
4 - Heat the remaining butter in a frying pan, and make your pancakes.
5 - spread your pancakes with marmalade, fold in half and sprinkle with icing sugar. Grill for 2 minutes.
6 - Arrange your pancakes on a plate and drizzle with the orange mixture.

GOOD POINTS – looked attractive and presentable.
Batter was even.
Sauce was smooth.
BAD POINTS – sauce was too tangy.

IMPROVEMENTS NEEDED – sauce could be made sweeter and possibly slightly thicker

WORTH CONSIDERING DEVELOPMENT? Possibly as could be modified to be better and can be made in the time that we are given. Also there are lots of different things I could change with it to choose the best.

Hedonic rating test

	Score	Taster 1	Taster 2	Taster 3
Like a lot	5			
Like a little	4			4
Neither like nor dislike	3	3	3	
Dislike a little	2			
Dislike a lot	1			
TOTAL :	11 /15			

Overall Evaluation
This product looked attractive but the sauce did not have a very nice taste. The batter itself was the right shape and was cooked for the right amount of time therefore was the right colour. The sensory profile shows that the product matches the desired profile with the appearance but not with the taste.

B *Presenting design ideas*

Activity

Identify from the following list of ideas which ones you would choose to make with reasons.

Lasagne, apple pie, profiteroles, cheesecake, fish cakes, lemon meringue pie, cauliflower cheese.

Summary

You should now be able to:

make good quality products which people would choose to eat

record what you have done.

13.1 Sensory testing methods

■ What is sensory analysis?

The enjoyment of eating is affected by what food looks like, tastes like, feels like and smells like. These are sensory qualities. Sensory testing of these qualities is carried out throughout the development of a food product and the results used to improve or change it. We call this **sensory analysis**.

■ Sensory testing methods

Sensory testing can be carried out by trained or untrained testers.

Ranking tests

- These type of tests are used to test similar products in terms of a specific flavour, e.g. sweetness.
- Each sample should be coded and not put in a rank order. There should be a minimum of ten untrained tasters.
- Testers would need to put the products in order of sweetness.
- Results could be recorded on a table similar to the one below.

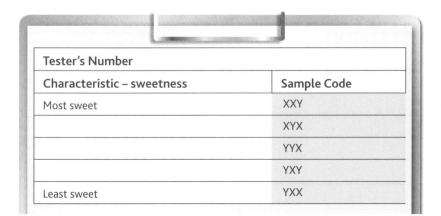

Tester's Number	
Characteristic – sweetness	**Sample Code**
Most sweet	XXY
	XYX
	YYX
	YXY
Least sweet	YXX

In your work you might use this type of test if you were trying to reduce the sweetness of a product and you gave testers samples from different batches that you had made.

Profiling tests

The most common one which you will use is the product profile recorded as a star diagram. It is normal to use either a six point star

A Sensory qualities

B Coded samples

depending on the sensory characteristics you need to profile. You need to use at least ten trained testers who are given coded samples. Each tester rates each characteristic on a scale of 6 (1 being the lowest and 6 the highest). Results for each tester are totalled and averaged and it is this score which is put onto the star diagram to give a visual profile.

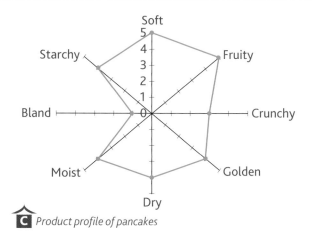

C *Product profile of pancakes*

Rating tests

These tests are used to assess a specific flavour or texture but require trained testers and the use of a particular scale. This could be a five-point or seven-point scale when testing the overall sensory characteristics of soup.

Tomato Soup – seven point scale

1 Dislike extremely.

2 Dislike a lot.

3 Dislike a little.

4 Neither like or dislike.

5 Like a little.

6 Like a lot.

7 Like extremely.

Rating tests can also be used for one particular attribute, e.g. saltiness, sourness.

Difference tests

These tests are used to find out if there is a clear difference between products. They might be used to test a low fat product against the same product with a higher fat content or the flavour of a traditional breakfast cereal with an economy brand. Often samples will include two identical products and one different product or three identical ones and two different.

You might use this test if you are developing your product by changing one ingredient in each development to find out if people can notice any differences.

Activity

Use difference testing to compare four different types of milk for flavour.

Key terms

Sensory analysis: identifying the sensory characteristics of products, i.e. taste, texture, appearance, mouth-feel, colour.

AQA *Examiner's tip*

■ When you are asked to describe a method of sensory testing ensure you give clear details about how to set it up, carry it out, record the results and draw conclusions.

■ Always use at least ten testers.

Summary

You should now be able to:

explain the meaning of sensory analysis and why it is used

describe different methods of sensory testing.

⬭links

Visit the BNF website – **www.bnf.org.uk**.

13.2 Using sensory testing

When will I use sensory testing?

Sensory analysis is an important part of product development. A product's sensory qualities are affected by the types, qualities, proportions and quantities of the ingredients used plus consumer preferences for particular sensory **attributes**. You will find sensory testing is important at the following points in the design process:

- at the research stage when you are analysing existing products or making products of your own
- at the design idea stage when you are making decisions about selecting which product to develop
- at the development stage when you are adapting a product.

Which methods of sensory testing should I use?

You need to understand all methods of sensory testing as these could be tested in the written examination. For your controlled assessment you need to use ones which are appropriate.

How do I select which methods to use?

Always have an aim.

Examples of aims:

- to find out which type of fruit people prefer in fruit crumbles
- to find out if curry sauce meets its sensory specification
- to find out if people prefer whole wheat or spinach flavoured lasagne.

The methods you use depend on what you need to find out. Example number one requires a preference test, number two a profiling test and number three a difference test.

How do I carry out sensory testing in the classroom?

In industry specialist tasting booths are used.

In your classroom you may not have a special area for sensory testing but always:

- invite at least ten testers to taste your food products (it is better to use strangers rather than your best friends as you should get a more honest answer)
- set up an area in the classroom which is clean, quiet and as far as possible free from smells
- use the right equipment – see picture below
- use the right **descriptors**.

Objectives

Understand when to carry out sensory testing.

Understand how you will carry out sensory testing.

Understand how you will record and use results of sensory testing.

Remember

Remember that tasters need to use appropriate descriptors and not words like 'nice'.

Remember

If a product does not meet its sensory profile you need to make it and remake it until its desired characteristics have been achieved. Also if people decide on particular preferences you need to develop your product to meet these.

Key terms

Attribute: particular characteristic of food, e.g. crumbly texture.

Descriptor: a word describing a sensory characteristic, e.g. spicy.

Sample: small amount of the product.

A *Tasting products in the classroom*

- Ensure that **samples** are the same size, served on a white plate.
- Code each sample.
- Provide water or lime juice for tasters to drink between tasting samples.
- Have a ready prepared recording sheet.

⚭links

See 19.2 Sensory testing on page 166.

B *Tasting food*

Using results

The most important part of sensory testing is knowing how to use results. It is important that you analyse results to find out likes, dislikes and things to improve.

(Activity box)

Activity

Produce a checklist to be displayed in the sensory testing area of your classroom explaining how to carry out a rating test.

AQA *Examiner's tip*

For written exams you need to give very full explanations of methods and/or how to use them.

Always use appropriate vocabulary to describe sensory characteristics.

Summary

You should now be able to:

choose appropriate methods of sensory testing

carry out successful sensory testing in the classroom

know how to record and use results.

⚭links

See Controlled Assessment pages 160–161.

kerboodle!

14 Making decisions

14.1 Evaluating design ideas

How to evaluate design ideas

When you have made a number of your chosen design ideas you need to evaluate these in order to select one you can develop. It is important that you look back at the design criteria you wrote and use these for your evaluation. When you are evaluating your ideas you should give your aim, e.g. I am aiming to choose a product which will enable me to carry out a lot of **development** to show my skill and also produce an original product made to a high standard.

The example below shows you one way of evaluating your design ideas.

Objectives

Understand how to evaluate design ideas against design criteria.

Understand how to select and idea for development.

∞ **links**

See 19.1 Evaluation of design ideas.

In order to decide which idea to develop I need to evaluate all my ideas against the Design Criteria. The table below show the criteria and the product number.

For each idea marks out of 5 can be given.

Design criteria Product Number	1	2	3	4	5	6	7	8
Essential criteria								
Colourful	3	3	4	1	4	4	4	5
Spicy	5	2	4	2	3	5	3	4
Sold chilled	5	5	5	4	5	5	5	5
Serves 2	3	3	4	3	5	4	5	4
Cost no more than £3.99	4	5	5	3	5	4	4	5
No artificial preservatives, flavours and colours	5	5	5	5	5	5	5	5
Must be a savoury Mexican main course	5	3	4	5	5	3	5	5
Desirable criteria								
Grilled or oven cooked	1	0	5	5	3	1	1	5
Contain vegetables	5	5	3	4	2	5	5	5
Suitable for vegetarians	5	5	1	5	1	1	5	5
Total	41	36	40	37	37	37	42	48

The idea which best fits the design criteria is number 8 – Vegetable Fajitas

A *Evaluation of design ideas*

There are other ways of evaluating design ideas:

In addition to the information which you collect from the **evaluation** of your ideas you need to think about products that will lend themselves to a good amount of development. The following are examples of product which could be developed in a range of different ways.

Remember

It is recommended that you only choose one idea to develop as this will enable you to do lots of development work.

Possible ways of developing ideas

The examples below give you a starting point for considering which idea is better to develop. You need to think about a product which would be able to be changed in terms of:

- different aspects of a product, e.g. one with a casing and filling or one which is layered
- types of ingredients/ alternative ingredients
- methods of making
- shape and size
- nutritional content
- cooking methods
- flavours, textures
- cost
- portion size
- finishing techniques.

B *A food product which could be developed in a range of different ways*

> ### Key terms
>
> **Development**: make changes to a food product which will affect its characteristics.
>
> **Evaluation**: summarise information and make conclusions, judgements and decisions.

Base:
Different meats, vegetables, meat alternatives
Different types of basic sauces – tomato, pesto, brown
Addition of herbs and spices

Pasta:
Different flavoured lasagne – spinach tomato, plain
Use of different types of pasta shapes including cannelloni
Use of pre cooked, dried uncooked or fresh
Making own pasta

Lasagne

Size and storage:
Different portion sizes
Testing effects of chilling and freezing

White sauce:
Different methods and milks and consistencies
Addition of different cheeses
Addition of other ingredients, e.g. herbs
Addition of different sized vegetables to vary textures
Different layering of sauces

Functions of ingredients:
Investigations and experiments with small samples

Toppings:
Use of cheese, bread-crumbs, cereals, tomatoes, peppers, etc. to improve appearance and texture

C *Development ideas for products*

> ### AQA Examiner's tip
>
> - In a written examination you need to be able to explain how to evaluate design ideas.
> - In controlled assessment and in the written exam you need to show ways in which design ideas can be developed. These should be detailed, clear and imaginative.

> ### Activities
>
> 1. Suggest developments for the following products:
>   - Shepherd's pie ■ muffins ■ pancakes.
> 2. Think about a favourite food product and suggest ways it can be developed.

> ### Summary
>
> You should now be able to:
>
> write an aim for your evaluation
>
> show how ideas meet or do not meet design criteria
>
> suggest ways of developing ideas.

14.2 Product analysis

What is product analysis and why is it used?

Product analysis means examining an existing product to find out information about its characteristics and those of its packaging.

Product analysis is carried out to find:

- how the product is made
- what ingredients have been used
- how these ingredients have been combined to achieve the structure and characteristics of the product
- its **nutritional content**
- its sensory attributes
- its cost
- shape and size
- storage and shelf life
- the type of packaging that is used
- how and what information is provided by the packaging.

Product analysis can be used at different stages in the design process. Sometimes a product(s) could be analysed when you are doing your research. It is very useful to look at a similar product before you begin your development, e.g. if you have decided to develop a savoury flan you could buy a ready-made quiche to analyse. This could inspire you with developments for your chosen product.

A *A selection of ready-made products*

Carrying out product analysis

The example below shows the type of information you can get from a product analysis.

The following information was obtained from the analysis of a pizza:

'Thin white bread base, covered with tomato purée, topped with mozzarella cheese and slices of pepperoni slotted into the cheese.

B *A product and its packaging*

Circular in shape, weighs 225g, keep refrigerated, long shelf life, can be frozen, serves two, cost £2, suitable for people on limited budget, e.g. students and low income families.

Attractive appearance when cooked. Good range of textures, limited range of flavours and colours.

Good range of information on packaging.

OVERALL EVALUATION: Good value for money, fairly ordinary flavour, needs more colour and addition of other ingredients. Base was bland and could have herbs or cheese added. Base could be made using different flours or from scone dough to give a different texture. Stuffed crust or folding dough would add interest. Good idea to slot slices of pepperoni in the cheese to help distribution.'

Product specification

A **product specification** gives details of a product's characteristics including shape, size and sensory qualities.

A **test kitchen** works to a product specification in order to produce a consistent product. The pizza will have been made to a product specification.

You will need to write a product specification before you begin to develop your product in order to achieve desired characteristics and meet the requirements of your design brief. The following **mnemonic** can help you:

S Shape

A Appearance

T Taste and texture

S Size

U Unit cost

M Materials

A Age(target group)

S Storage

Summary

You should now be able to:

choose and carry out a relevant product to analyse

understand how the product's characteristics have been achieved

write a product specification.

∞ links

Visit British Nutrition Foundation **www.nutrition.org.uk**.

Visit Sainsbury's Website **www.sainsburys.co.uk**.

Key terms

Product analysis: examining a food product to find out the ingredients, packaging characteristics and properties.

Nutritional content: the type and quantity of nutrients which the product supplies.

Product specification: a list of features/characteristics/properties which a food product must meet.

Test kitchen: the place where a food technologist experiments and develops new products.

Mnemonic: something to help us remember things.

15 Product development

15.1 Development: development methods

What is development?

New products arise as a result of food technologists and manufacturers taking an existing idea or product and changing it in a range of different ways to make a new prototype. Development involves:

- using different ingredients and amounts
- making mixtures using different methods
- changing the nutritional profile
- changing sensory characteristics
- testing different storage methods and effects on shelf life
- creating new shapes
- altering portion sizes
- meeting a cost need
- meeting a dietary need.

Development is a complex process. The product will be affected by changes in:

- composition
- nutritional content
- structure
- effects of storage.
- sensory profile

Development will be a key part of your Controlled Assessment.

Based on these results the final product would be made which took into account the original product specification and the effects of the changes.

Functions of ingredients

The characteristics of a food outcome are achieved by ingredients used in different ratios and proportions. For example:

- rubbed-in mixtures
- using half fat to flour and no liquid (a crumble)
- using 2/3 fat to flour (shortbread)
- using half fat to flour and water and stiff dough (short crust pastry)

Name of product   Victoria sandwich cake

Recipe   100g butter
100g caster sugar
100g SR flour
2 medium eggs
2 tbsp raspberry jam

Objectives

Understand the difference between development and modification.

Understand how to use different methods, ingredients and ratios and proportions to develop a product.

Remember

Sensory evaluation should be carried out after each development or modification.

AQA **Examiner's tip**

- Development activities gain more credit than modifications.
- You need to be able to describe, evaluate and make conclusions from development activities in both controlled assessment and the written exam.

links

See Chapter 18 Development and making of design proposals.

A   *Victoria sponge*

B *Example of how a Victoria sandwich cake could be developed*

Development	Effect	Development	Effect	
Developing the fat		Developing the flour and method		
Use block margarine	Poorer flavour Same nutritional content High in saturated fat	Use wholemeal flour and baking powder	Less rise Dense texture Poor flavour and mouthfeel	
Change margarine to polyunsaturated fat	Same fat content but unsaturated Poorer flavour	Increase amount of BP	Better rise and texture Poor flavour	
Use low fat spread	Less rise Close texture Poor flavour and colour	Use all in one method for original recipe	Good flavour and texture Insufficient rise	
Reduce amount of fat	Poor texture Insufficient flavour and rise Shorter shelf life	Add small amount of baking powder	Better rise Texture slightly more dense than creaming	
Developing the sugar		Developing the flavour and shape		
Reduce amount of sugar	Poor rise Dense texture Flavour impaired	Make up original mixture and divide into four – flavour each sample a) cocoa b) coffee c) lemon d) coconut	a) Dryer mixture, cake rises to a peak b) Soft consistency, less rise c) Good consistency and rise d) Dry consistency, cake rises to peak	
Use sugar alternative	Poor texture Good rise Bland flavour and aftertaste Expensive	Bake in one square tin	Rises to a peak Centre undercooked Temperature too high	
Use soft brown sugar	Poor colour Open texture Less rise Same sugar content	Bake in swiss roll tin	Dry, crisp texture Overcooked Temperature too high	

What is modification?

Modification is a simple process whereby only limited changes are made to a product.

Modifications are usually substitutions which have little or no effect on structure, texture, shape and appearance of a product. Flavour might be changed. Little or no objective evaluation is done on the changes and product is made on personal choice.

C *Example of how a Shepherd's Pie could be modified*

Modifications	Effects on product
Lamb changed to beef	Different flavour
Red onion used instead of white	No change
Potatoes mashed with margarine	Slightly less flavour
TVP used instead of meat	Poor flavour

Key terms

Modification: simple changes which have little effect on the structure and composition.

Summary

You should now be able to:

describe the difference between development and modification

understand functions of ingredients

develop a product by changing ingredients, methods and proportions.

kerboodle!

Development: nutritional

What is nutritional analysis?

The link between diet and health is a major concern within the UK. Obesity in children, heart disease, diabetes, allergies, and ingredients used in food products are issues which food manufacturers have to address.

The nutritional content of a product is therefore very important to the consumer.

Sometimes products are designed with a specific nutritional focus, e.g. low fat, low sugar, and high fibre. Also some products are marketed as providing a high daily percentage of a specific nutrient. You might have included one or more of these points in your product specification. Therefore as you begin to develop your **prototype** you will need to keep a check on its nutritional content.

How do I check the nutritional content of my products and prototypes?

You are able to check the nutritional content of a product by modelling. This means taking a recipe and by using a computer programme called Nutritional Analysis Software the product's nutritional content can be found. When you are developing your prototype, modelling the recipe before making it cuts down on cost and also gives you evidence for your folder about why you have made certain decisions.

The examples below, which link to the development of a low fat product, show the nutritional content of both an existing recipe and a developed recipe.

Conclusions

Sometimes by changing the nutritional content the sensory characteristics of a product are changed, e.g. by reducing sugar and fat the flavour and texture of a product will be affected. In the pizza the fat content has been reduced and in particular there is a reduction in saturated fat which meets the product specification; however by using low fat cheese it is likely that the flavour and texture of the pizza might be affected and in particular the cheese will not melt to the same extent.

Key terms

Nutritional analysis: using resources (e.g. a computer) to find out the nutritional content of a product.

Prototype: the final food product developed by the test kitchen.

⬯links

Visit British Nutrition Foundation website at **www.nutrition.org.uk**

See Chapter 10 Labelling and packaging.

Activity

Use nutritional analysis software to compare an original recipe for cheese biscuits with a developed recipe. Produce a report to a food manufacturer showing how you have improved the nutritional content and how this has affected the sensory characteristics.

Nutrition information - Original Pizza		
	Typical values	
	per 100 g	per serving 208 g
Energy	1033 kJ 247 kcal	2148 kJ 513 kcal
Energy	9.29 g	19.3 g
Carbohydrate of which sugar	30.7 g 3.22 g	63.9 g 6.7 g
Fat of which Saturates	10.5 g 6.0g	21.8 g 12.5 g
Fibre (NSP)	1.42 g	2.95 g
Sodium	0.26 g	0.53 g
Iron	1.09 mg	2.27 mg
Calcium	164 mg	341 mg
Vitamin A	117 mg	242 ug
Thiamin (B1)	0.22 mg	0.45 mg
Riboflavin	0.13 mg	0.28 mg
Niacin	1.49 mg	3.1 mg
Vitamin C	4.96 mg	10.3 mg
Vitamin D	0.07 mg	0.14 ug

Dietary Reference Values

% Energy - Protein	15
% Energy - Fat	38
% Energy - Carbohydrate	17

DRV based on requirements for a boy aged 11-14 years

DRV >|< 100%

% Energy	139
% Protein	275
% Sodium	199
% Calcium	204
% Iron	121
% Vitamin	242
% Vitamin	300
% Vitamin	138
% Niacin	124
% Vitamin C	177
% Vitamin D	*****************

Nutrition information - Amended Pizza		
	Typical values	
	per 100 g	per serving 200 g
Energy	845 kJ 202 kcal	1691 kJ 404 kcal
Energy	10.1 g	20.3 g
Carbohydrate of which sugar	32.1 g 3.4 g	64.1 g 6.8 g
Fat of which Saturates	4.57 g 1.78 g	9.14 g 3.56 g
Fibre (NSP)	1.48 g	2.95 g
Sodium	0.23 g	0.47 g
Iron	1.13 mg	2.26 mg
Calcium	186 mg	372 mg
Vitamin A	70.4 mg	141 ug
Thiamin (B1)	0.2 mg	0.41 mg
Riboflavin	0.16 mg	0.32 mg
Niacin	1.45 mg	2.9 mg
Vitamin C	5.17 mg	10.3 mg
Vitamin D	0.18 mg	0.36 ug

Dietary Reference Values

% Energy - Protein	20
% Energy - Fat	20
% Energy - Carbohydrate	60

DRV based on requirements for a boy aged 11-14 years

DRV >|< 100%

% Energy	109
% Protein	288
% Sodium	174
% Calcium	223
% Iron	120
% Vitamin	141
% Vitamin	271
% Vitamin	158
% Niacin	116
% Vitamin C	177
% Vitamin D	*****************

A *Nutritional analysis of pizza*

Other ways to develop prototypes

As you work your way through the development of your prototype you could consider the following types of development methods:

- changing shape
- changing size – this will have an effect on the nutritional content of a portion
- testing out different storage methods and their impact on shelf life (a product may be unsuitable for chilling and/or freezing as its texture and flavour may be impaired by the process).

AQA Examiner's tip

- In your coursework you need to be able to contrast and compare results from nutritional analysis, drawing conclusions which inform the development process.

- In the written exam you are likely to have to comment on the results from nutritional analysis.

Summary

You should now be able to:

use nutritional analysis to analyse products and model development ideas.

Remember

You may have to carry out nutritional analysis when you make your design ideas if your design criteria include a nutritional aspect.

Investigating and experimenting during development

Complete products do not need to be made each time a development or modification is carried out. In a test kitchen the food technologist will be working with small samples of ingredients and foods to investigate possible developments for a food prototype. This saves time and money.

As you develop your prototype you can do the same. Therefore it would be possible to carry out several development activities in one lesson.

What is fair testing?

In order to make results reliable and comparisons equal you need to ensure that each time you carry out an investigation or experiment you do it fairly. This means you compare 'like with like', only have one variable each time and have a control sample against which you should evaluate your results. The following example shows you how to do it.

Investigating the effect of different fats and flours in short crust pastry:

Control sample: 50g plain flour 2 tsp cold water

Sample number	Addition
1	12½ g block margarine and 12½ lard
2	25g lard
3	25g block margarine
4	25g butter
5	25g low fat spread
6	25g soft polyunsaturated margarine
7	25g olive oil
8	25g white vegetable fat
9	12½ block margarine
10	12½ white vegetable fat

All samples can be baked on one tray in the same position in the oven and at the same temperature.

The only variable is the fat and the results can be compared with the control sample. In the light of the results an assessment of how far the pastry meets the product specification can be carried out together with its sensory characteristics and its fitness for purpose. You are likely to be commenting upon the pastry's:

- texture
- flavour
- colour before and after cooking
- general palatability
- suitability for rolling and shaping

A Food technologist in the test kitchen

B Pastry samples

- ability to keep shape during cooking
- nutritive content
- quantity of water needed.

Once you have evaluated each sample and made your decision about which is the most suitable you could carry out further investigations by adding other ingredients to the basic mixture, e.g. cheese, herbs, colourings, nuts, grains, seeds, etc. You could also experiment with other types of flour.

Experimenting with small quantities

You could make one mixture and divide into smaller samples, such as:

- making small buns rather than a large cake
- adding different ingredients to samples of white sauce
- adding ingredients to scones to give different flavours and textures
- adding ingredients to bread mixtures
- using small tartlets and different fillings rather than re-making a large flan
- making small versions of sausage rolls and pasties with different fillings and possibly different glazes
- using small, ready made biscuits or small buns to trial icings and toppings
- making small pies in bun tins with different fillings
- making a batch of pasta and add different flavourings/ingredients. Make different fillings and test shapes.

C *Products suitable for developing in small quantities*

Summary

You should now be able to:

carry out investigations and experiments using small quantities

apply fair testing to your work

use results to inform the next steps in the process.

Key terms

Fair testing: to compare like with like using only one variable.

⊙⊙links

See Controlled Assessment pages 140–43.

See 13.1 Sensory testing methods on page 110.

AQA　*Examiner's tip*

- It is essential when testing your own developments that you use people in your target group not just friends.
- In the written exam you may be given results from experiments or investigations and be asked to comment on these and make conclusions with reasons about which might be the best to take forward.

Remember

You should always have an aim for each development – this shows your thought process and gives you something to evaluate your results against.

Activity

You are trying to develop a new filling for ravioli using a tomato-based sauce. How could you develop a range of different fillings by starting with 250ml of basic tomato sauce?

Why is sensory evaluation so important?

In order to meet a product specification each development needs evaluating. As sensory characteristics are one of the most important aspects of a food product, sensory analysis should form the main part of the evaluation.

Objectives

Understand how to use the results of sensory testing to improve a product.

Understand how to evaluate developments against a product specification.

Using the product specification

The following example is a product specification for samosas.

- Interesting shape.
- High fibre.
- Crisp outside texture.
- Spicy/fruity filling.
- Golden colour on outside.
- Easy to eat by hand.
- Small individual pieces of vegetable, fruit or meat in filling.
- Range of textures in filling.
- Soft but not runny filling.
- Good contrast between filling and casing.
- Contributes to 'five a day'.

An existing recipe has been used as a starting point for developing the samosas. A panel of tasters has been asked to analyse the outcome from the recipe by using a rating chart.

A Samosas

	1	2	3	4	5	6	7	8	9	10
Dislike a lot			➥							
Dislike a little	➥					➥		➥		
Neither like nor dislike		➥		➥			➥	➥	➥	
Like a little					➥					➥
Like a lot		☆			☆				☆	

This gives a starting point for development and also the results show there is lots of opportunity to develop the samosas.

After several more developments another panel of five tasters carry out a **Profiling Test** using a star profile to evaluate eight specific characteristics of the product. The results are shown below:

These results indicate that the crispness and the texture of the pastry need to be improved together with the texture and taste of the filling. This is also providing some evaluation against the product specification.

At this point it would be a good idea to carry out some small-scale development work using small quantities and small samples as described on page 122.

It would be useful to carry out Difference Tests at this point in order to find out if tasters can identify the changes you have made to the original samosa recipe. Two samples could be made from the original recipe and one from an improved recipe. The results could be used to make decisions about developing the samosas further.

Key term

Profiling test: a sensory evaluation test to identify individual specific characteristics of a product.

∞ links

See 13.1 Sensory testing methods page 110.

See 19.2 Sensory testing page 166.

Activity

Using the results from the star profile, identify which aspects of the samosas need developing further and suggest ways in which you could do this.

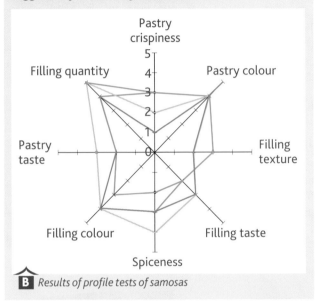

B *Results of profile tests of samosas*

AQA *Examiner's tip*

■ Carry out regular sensory testing of your development work.

■ In the written examination make clear, detailed suggestions for improvements and if necessary explain how you would carry them out.

Remember

■ Sensory testing has to be done fairly, correctly and hygienically.

■ You must always prepare recording sheets for your testers to use.

C *Example of a Difference Test for samosas*

Summary

You should now be able to:

use testing to improve your product as it develops

check that the sensory attributes of your product meet the product specification.

16 The final design solution

16.1 Reviewing the process

■ The final product

How to show good making skills

You will have provided evidence of 'making skills' throughout your project. These will include:

- making ideas at the 'research stage'
- analysing a product
- making a range of design ideas which show different making skills
- carrying out sensory testing
- demonstrating a range of development skills which include product appraisal, investigations, experiments, testing
- evaluating the functions of ingredients and the impact that this has on development, e.g. by reducing the fat in a roux sauce the result demonstrates poor flavour and less gloss; by reducing the sugar in a cake there is less rise, poor texture and poor flavour
- making your final product.

Standard components

You could re-make your product using **standard components** – a manufacturer would not make each part of the product from scratch. Many components are 'bought in', e.g. a fruit pie would be made using either chilled or frozen pastry, or a ready-made pastry case, a pre-prepared pie filling and a glaze which is already mixed using pasteurised egg.

Achieving a quality product with a high finish

- Weighing and measuring accurately.
- Storing your food correctly – **high risk food** needs to be stored at the correct temperature, lower risk food including all fats needs to be refrigerated otherwise they will melt and their function is impaired.
- Understanding the recipe and method well – if you are making a product for the first time you may not achieve the desired finish.
- Making products with which you are confident.
- Using the correct cooking method, oven temperature, shelf position and hob control.
- Using equipment, e.g. food mixer to ensure **consistency**.
- Knowing how much mixture is required in the baking tin, casserole or pan.
- Using the right amount of fillings for different casings, e.g. flans, pasties, cakes.

Objectives

Understand how to use a range of making skills.

Understand how to produce a final product of high quality and finish.

Understand how to present information about the final product.

Understand how to write a short evaluation against the original design brief and design criteria.

Understand how to design a production plan.

A *Examples of standard components*

∞links

See Chapter 5.1 page 56 for more detail on standard components.

- Applying **quality control** to achieve uniformity and consistency.
- Judging when products are cooked/set – not over or under.
- Allowing cakes/desserts to cool/set before filling/icing.
- Practising decorating and finishing techniques such a icing, piping, applying decorative edges to pastry, shaping bread and scones.
- Having pride in your work and making products which people **WANT** to eat.

B *Techniques, processes and equipment which will assist in the making of high quality products*

Presenting your information

You need to be able to present appropriate evidence of how you have achieved your final product. This can be done by:

- summarising the developments you have made with reasons
- providing a recipe for your prototype

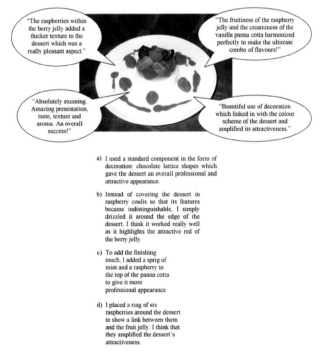

"The raspberries within the berry jelly added a thicker texture to the dessert which was a really pleasant aspect."

"The fruitiness of the raspberry jelly and the creaminess of the vanilla panna cotta harmonized perfectly to make the ultimate combo of flavours!"

"Absolutely stunning. Amazing presentation, taste, texture and aroma. An overall success!"

"Beautiful use of decoration which linked in with the colour scheme of the dessert and amplified its attractiveness."

a) I used a standard component in the form of decoration: chocolate lattice shapes which gave the dessert an overall professional and attractive appearance.

b) Instead of covering the dessert in raspberry coulis so that its features became indistinguishable, I simply drizzled it around the edge of the dessert. I think it worked really well as it highlights the attractive red of the berry jelly.

c) To add the finishing touch, I added a sprig of mint and a raspberry to the top of the panna cotta to give it more professional appearance

d) I placed a ring of six raspberries around the dessert to show a link between them and the fruit jelly. I think that they amplified the dessert's attractiveness.

Functions of Ingredients:
The panna cotta contains large quantities of milk and cream which provide the bulk of the mixture and give it the rich, creamy flavour and the velvety, soft texture. The milk and cream, when hot, also provide the heat and liquid needed for the gelatine to dissolve. The vanilla essence is the main source of flavour in the panna cotta and can be very strong if too much is added, however 1 teaspoon is just enough to lift the flavour of the panna cotta to make it delicious and refreshing when combined with the fruitiness of the raspberry jelly. The gelatine is what gives the dessert its soft, gelatinous structure and ensures that the panna cotta can remain upright without collapsing.
In the berry jelly, the cranberry juice is what gives it that rich, ruby red colouring and the sweet fruity flavour which goes so will with the creaminess of the panna cotta. The lemon juice supplies the jelly with the extra tartness which the cranberry juice cannot provide. The raspberries are the main bulk ingredient of the berry jelly, as they add colour and texture to it by giving the consumer something to chew whilst eating. The sugar sweetens the overall taste, and draws out the fructose of the raspberries so that the jelly is sweetened by not only the juices and sugar, but the fruit aswell.

C *Example of some inclusions for a final review*

- carrying out a final nutritional analysis
- explaining the functions of ingredients in the final product
- providing results and conclusions from sensory evaluation of the final product
- including an annotated photograph of the final prototype
- comparison with original design criteria
- evaluation against design brief.

Planning for production

In order to make your final product in the test kitchen the food technologist needs a production plan which gives details about steps in the process, quality controls, hygiene and safety checks. There are several ways a plan can be represented, like these shown on page 129.

D Steps in the process of making a chicken and leek pie

A table

Sequence	Quality control	Hygiene and safety
The order and specific timings should be included here	Ways in which the product's quality can be ensured go here	Aspects related to food storage and hygiene in the test kitchen go here
E.g. cut chicken into small pieces ready to be fried	Ensure chicken is cut into even sized pieces	Remove chicken from refrigerator immediately prior to preparation Use appropriate cutting board to prevent cross contamination Wash hands immediately after preparing the chicken, put cutting board to be washed at high temperature

A flow diagram

Symbols showing the steps in the making process with points where decisions need to be made are a good way of designing a production plan. The following symbols are those used generally:

 Process symbol

 Decision symbol

 Alternative process symbol

E *Ways of presenting a production plan*

Remember

You are only producing a plan for a prototype in the test kitchen not for mass production in a factory.

Activities

1. Make a batch of scones to show how you have applied quality control.

2. Design a production plan for making a chicken and mushroom pie. Include quality controls and hygiene and safety requirements.

3. Imagine you are to present your idea to the product development manager of a large food manufacturer. You are only able to take one sheet of paper into the presentation. How would you present your ideas?

⊙⊙ **links**

See Chapter 3 Combining ingredients.

AQA *Examiner's tip*

- Credit will be gained for showing higher level making skills carried out to a high standard.
- Give sketches and detailed explanations in the written examination if asked about how a product could be finished.
- You need to be concise when designing a production plan for your own product.
- In the written examination you may be asked to design a production plan for making a given product. Therefore you need to ensure you learn how to create a step-by-step plan using flow diagrams or charts.

Summary

You should now be able to:

demonstrate a range of making skills

make high quality products which show skill and expertise

use all the information from your developments to produce a food product of high quality and finish

select and use equipment which ensures quality control

meet the product specification and the design criteria.

kerboodle!

Examination-style questions

Questions will require you to show knowledge and understanding of the working characteristics of food together with processing techniques in order to design, make and evaluate food products which meet health, dietary, socio economic and cultural/religious needs of different groups within our society.

1   Sales of bread products have changed over recent years.
The table below shows sales figures for different breads.

	1960s	1980s	2000
White bread	75%	50%	41%
Ciabatta	8%	20%	24%
Croissant	5%	11%	12%
Naan	10%	12%	15%

(a) (i)   Name three different research methods that manufacturers may use to gather this type of information. *(3 marks)*

(ii) Explain why white bread has become less popular. *(4 marks)*

(b) (i)   How do manufacturers of sandwich products ensure they meet the needs of consumers who follow special diets for health reasons? *(6 marks)*

(c) (i)   Explain how manufacturers could use the internet when designing and making a new range of food products. *(4 marks)*

(ii) What other types of electronic media may be used when researching existing products? *(2 marks)*

2   This is a recipe for fruit slices.
Ingredients:
150g wholemeal flour
100g white flour
50g soft brown sugar
125g polyunsaturated margarine
200g apricots

(a) (i)   Explain why manufacturers use sensory testing. *(4 marks)*

(b)   The profile below shows the results of sensory testing on fruit slices.

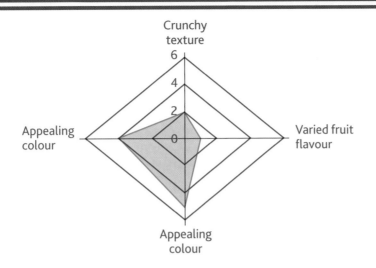

(i) Evaluate the effectiveness of this food product.
Identify where improvements may be made. *(6 marks)*

(ii) Suggest what actions the manufacturer may carry out to improve this product. *(6 marks)*

(iii) Manufacturers often use symbols to pass information onto consumers.
The following symbol is shown on the packaging of the fruit slices.
What does it mean? *(3 marks)*

(c) Explain how manufacturers could develop food products that are suitable
to meet consumers' need for environmentally friendly products. *(6 marks)*

 These questions are based on general food products but similar questions may be asked on any chosen food product. You may like to practise similar questions for the food product of your choice.

Introduction

 What is design and making practice?

A GCSE grade for Food Technology is achieved by completing two units:

- Unit 1   A written paper worth 40% of the total marks
- Unit 2   A project which is worth 60% of the total marks. This project will enable you to show your designing and making skills and therefore is called Design and Making Practice. It is assessed to criteria called 'Controlled Assessment Criteria' as shown below. The five criteria together are worth 90 marks.
  - Investigating the Design Context.
  - Development of Design Proposals.
  - Making.
  - Testing and Evaluation.
  - Communication.

Your designing and making project must meet strict guidelines, hence the name 'Controlled Assessment'.

A successful project will address all elements of the design process.

The following points give you more information about what you have to do:

- With the guidance of your teacher you must select a task from a given list of design which the examination board provides tasks.
- You must produce a design folder in which you record the process you have followed, the decisions you have made, reasons for making these decisions and evidence of the 'making you have done'.
- You must produce a final product prototype which would be similar to one produced in a test kitchen.
- You must produce a final design solution that could be presented to the manufacturer to make on a large scale.
- In addition to a final product you will make a series of other products at different stages in the process.
- All products, which you have made, will collectively provide evidence for your final 'making' grade. All your practical work will be assessed by your teacher throughout the project.
- Other activities will also contribute to the making grade. These include:
  - product analysis
  - sensory testing
  - investigative and experimental work.

- Your folder should consist of 20 sides of A3 paper, or equivalent A4 paper, or its ICT equivalent.
- You should spend a maximum of 45 hours on the project.
- Photographic evidence of the final product (and other products made) is required.

■ Carrying out your design and making practice

Your folder should follow the design process and be a concise record of what you have done in the process of designing and making your food product. It should also contain a record of what you have made and the quality of that making.

Some points to help you

- You need to plan your work carefully.
- Space on pages needs to be used well.
- The size of your writing/font size needs to be legible but not enormous.
- In particular you need to keep all research information concise.
- The folder should show your thought process and should use phrases which show why, how and what you did including an aim and conclusion for each stage.
- The quality of making needs to be very good. This means that your food products need to look appetising and attractive but also taste very good.
- In addition to making complete items you need to carry out development work which means you have to take an existing idea or recipe and keep changing it until you have achieved a new product. This will involve a lot of investigations and testing using small quantities.
- The functions of the ingredients you have used need to be understood and you need to comment on how these functions affect a product.
- Sensory testing should form an important part of your work.
- There should be some evidence of nutritional analysis.
- The quality of your written communication needs to be accurate.
- There should be evidence of some ICT within your project.
- Photographs of outcomes you have made and also the processes you have used are an excellent source of evidence for your 'making skills'.

How long should the project be?

The design folder should consist of approximately 20 sides of A3 paper or A4 equivalent or ICT equivalent. The space on each page should be used well and writing should be concise and accurate.

How much time should be spent on the project?

The complete project including all design work and all making work should take 45 hours.

Can I write my own design task?

No. You must use a task which has been set by the Awarding Body (AQA) and given to you by your teacher. You may be given several tasks to choose from.

What does 'target market' mean?

The target market is the specific or particular group of people at which your product is aimed. The target market might be related to nutritional needs or desirability, social factors including income/cost, age groups, luxury products, special occasions, e.g. a special diet, a value product, packed lunch product for a young child, a luxury dessert.

How much research do I need to produce?

Research can occur at any point in the design process and should always be relevant. Often a lot of research is done unnecessarily at the beginning of a project. A maximum of two sides of A3 at the beginning of your project is sufficient, supplemented by other research at relevant points.

Do I need to sketch my design ideas?

After a design specification has been finalised design ideas should be generated. This can be in the form of sketches, computer generated images, images from existing recipes or a combination of all of these methods. Lists of recipes must be avoided. When producing design ideas it is important to thoroughly annotate the ideas to show sensory properties, dimensions, working properties of ingredients and finishing techniques. Sketching is good preparation for the examination but not a mandatory requirement for the controlled assessment.

How many design ideas should I make?

There is no exact number but in order to demonstrate your making skills and provide evidence of a range of different skilful products you should make between four and six ideas.

How many products should I take to the developmental stage?

It is recommended that one product is taken to the developmental stage after carrying out detailed analysis and evaluation of your generated ideas. The initial product will undergo many changes at the development stage and this should be reflected in the amount of time spent on this section of the controlled assessment.

What is the difference between development and modification?

Development is more complex than modification and involves investigating and experimenting with ingredients and recipes. Development often has an impact on the structure, sensory qualities, nutritional profile and general acceptance of a food product. For example, using low fat spread in pastry may improve the nutritional profile but will impair texture, flavour and working characteristics.

Modification is a simple change where one ingredient is substituted for another, e.g. parsley instead of thyme.

Do I have to include a recipe for every product I make?

A method/process should not be included for each product or development activity that is carried out. The ingredients you are using should be documented and where relevant the working properties discussed. A detailed final recipe, including quality and food safety procedures, is required for the final design solution.

Do I need to make a complete product every time I cook?

It is not necessary to produce a complete product for all making activities. Complete products will be produced at the design ideas stage. However during development, investigation and experimental work, small samples should be tested to reduce costs and avoid wastage of ingredients.

How do I achieve high marks for making?

You must produce a range of products that use different skills, e.g. pastry making, sauce making. You need to show some originality, flair and innovation and make good use of development work to devise a final product. The products you make must have excellent sensory properties and be finished to a high standard. You need to be able to work independently and be organised throughout making activities. The final product must be accurately made and a quality item. Throughout making you need to show a thorough understanding of the ingredients that you are using. The use of standard components (ready made) ingredients will need to be avoided to attain high marks for making.

Do I need to make a final product?

Yes. After carrying out the design process your final product is the solution to the brief you have been set. The final product will result from all the development work you have carried out in the test kitchen.

Do I need to include photographic evidence?

Photographic evidence of the finished outcome must be produced. It is also strongly recommended that photographic evidence should be included at various stages of making. It is very good practice to show evidence of the completed design ideas and the different making stages of the final product.

Do I need to produce the packaging for my product?

You do not need to produce the packaging for the final product but a nutritional label should be produced. It is not necessary to discuss the packaging materials – this is an area to be tested on the examination paper.

Do I need to carry out nutritional analysis?

This is very much dependent upon the design brief you are working from. If the design brief has a nutritional slant, e.g. must be low in fat, then it is expected throughout the development of the product that the amount of fat in the products made will be considered. Once the final product has been made for any design brief, you should try to include nutritional information about the product on the food label.

Do I need to explain how my product will be made in a factory?

No. The coursework relates to what would happen in the test kitchen in the food manufacturing industry.

Do I get marks for good presentation of the folder?

It is important to present your work to the highest standard possible. There are six marks for good communication of your design folder. When marking your work your teacher will be looking for good spelling, punctuation and grammar. It is important to present your work in a clear and coherent manner. You will also be awarded marks for the use of technical language when explaining about the ingredients you are using.

Does my project need to be word processed?

No; however, criterion 5 assesses your written communication skills and word processing could ensure your work has text that is legible, easily understood and shows good spelling, punctuation and grammar.

Who will mark my work?

Your teacher will assess everything you have done which includes all work in your design folder and all the 'making' you have done. A moderator from the Awarding Body (AQA) will request some or all of the Food Technology work from your school and will re-mark it. You will them be awarded a mark for your Designing and Making Practice which will be added to the mark you get for the written examination. Your grade will then be calculated.

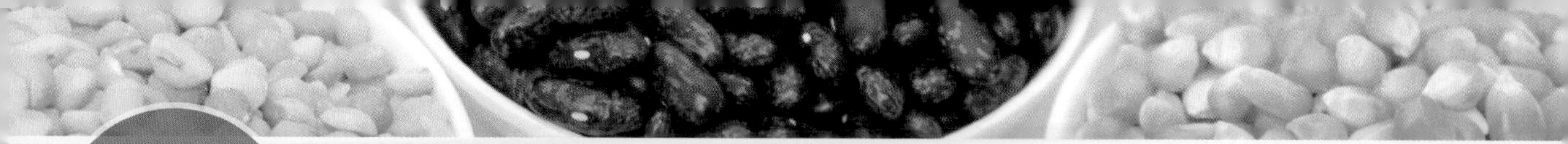

17 Investigating the design content

17.1 Research, analysis and criteria

Research and its analysis can take place at all stages of the design process as you feel the need to investigate further information about the product you are developing.

Select a design task from the specification. These are set in a context and enable focused research. This is an example of a context and design task.

Context

Context 1

Recent statistics have shown a rise in childhood obesity. A supermarket chain is responding to this by launching a food section aimed at 'low fat', 'low sugar' products for children.

Design task

Design and make a product suitable for sale in this section.

Context 2

Design task

The following are examples of some of the different types of research that can be carried out.

Jason's design brief is to make a new packed lunch product. He has analysed the brief thinking about the key words and thinking back to the development work he carried out in year 10 and has come up with the following plan for his research work, see Picture **A**.

Objectives

- Understand how to select relevant research methods to investigate the design brief.
- Understand how to carry out detailed research.
- Understand how to analyse research to write clear and specific design criteria.

Plan of Research

Before I can start to design ideas for my own packed lunch product, I will need to do some research to find information about the following things:

- Types of savoury packed lunch products already on sale.
- The size and shapes of products on sale
- Prices of savour packed lunch products on sale
- Different textures, flavours and fillings of savoury packed lunch products
- The different types of packaging used
- The different range of ingredients used in products
- The nutritional value of the products on sale
- The public opinion of savoury packed lunch products
- The methods of making and the recipes of savoury lunch products

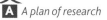

A *A plan of research*

Jason has carried out a questionnaire to ask a sample of people (ideally a target group if this is specified in the task) a range of questions. The questions must be written to ensure that the answers provide evidence to inform the design specification. It is not necessary to include the questionnaire. Ideally show the results of the questionnaire as a range of graphical images to show high level ICT skills, see Picture **B**.

AQA Examiner's comment

Jason has a wide range of research which he has decided to carry out to ensure that he has a variety of information that will enable him to write an informed design criteria to meet the brief.

Carrying out practical research is also a good way of including a range of making skills.

AC5 communication

Good use of bulleted points to get across concise information.

Ideas are communicated well with good use of accurate spelling.

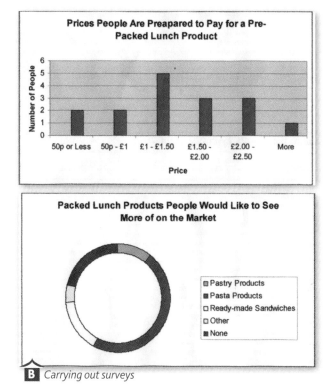

B *Carrying out surveys*

AC5 communication

Good use of two different types of graph to show results from the survey carried out and each graph is well labelled with a clear title.

AQA Examiner's comment

Jason uses a range of relevant graphs to show results from carrying out his survey on the target market. A detailed analysis of the results of the survey is then carried out to help write the design criteria.

A detailed product analysis helps the candidate look at existing similar products on the market to see a range of characteristics that could be included in the design specification for the new product. Product analysis looks at sensory characteristics as well as size, functional properties, packaging, storage and nutritional profile.

AC5 communication

- Excellent use of photographs and tables to explain results.
- Concise relevant descriptive words used to explain sensory characteristics.

Structure, dimensions, weight

Shape	Circular
Height	25 mm
Breadth	50 mm
Width	50 mm
Total weight	27 g
Ingredients	
Biscuit	20 g
Buttercream	4 g
Raspberry Jam	3 g

A descriptive taste test

Appearance	Neat, attractive, golden brown, appetising, well shaped biscuits, decorative, all identical.
Taste /Aroma	Fruity, sweet, buttery, floury, creamy.
Texture	Creamy, smooth centre, short, rich, melts in the mouth, crumbly, light.

C *Analysis of an existing product*

A good analysis of what the research has helped you to learn is essential to help you write your design specification. It is not an analysis of the research methods but the information researched.

Analysis of Research

From all of my research I have found that:

- There is a large variety of savoury pre-packed lunch box products already on the market. Examples include cold pasta salad, flavoured bread, pastry products (e.g. pasties, sausage rolls etc.) and various filled sandwich products (wraps, ready made sandwiches etc.)

- Prices range from twelve pence per serving to two pounds and fifty pence per serving – the large range is due to expense of ingredients used, and also the size of the product.

- People would like to see more pasta products (e.g. cold pasta salads) on sale. My questionnaire results tell me this.

- Lunch box products can be filled with a range of different products – these can be due to serving temperature – i.e. hot/cold. Hot products, e.g. pasties, are more commonly filled with vegetables and meat, whilst cold products, e.g. sandwiches, are more commonly filled with spreads and cold meat.

- A good, healthy lunch product should contain something from each of the food groups, and should contain all the nutrients necessary for the consumer. E.g. meat contains protein, and vegetables contain fibre – so a chicken and vegetable pasty would make a good product.

- The products are packaged in a variety of materials. These include plastic boxes, wrappers, plastic tubs, cardboard boxes and plastic trays. From my packaging research, I can see that the most commonly used type of packaging for packed lunch products is plastic, although it comes in may shapes and sizes.

- A packed lunch product should be easy to transport. It should be of a shape that is easy to store, and the product itself should be easy to eat without the necessary utensils. (Knife, fork etc.)

All the information I have collected throughout my research has been useful to me in some way, and will be taken into account whenever possible.

 **D** *Analysis of research*

Design criteria must be written using the new information evaluated from the analysis of research and may also consider the brief.

Design Criteria

I now need to produce a list of general criteria for my new packed lunch product, following my research findings.

My design ideas must:

- Be suitable for people of all ages, and be suitable for transport in small containers.
- Be well balanced and include the following nutrients: protein, carbohydrate, fat, vitamins, minerals and fibre.
- Cost between one pound and one pound and fifty pence.
- Look appetising and professional.
- Taste good – not taste burnt or stale etc.
- Have suitable packaging – strong, shape that is easy to store.
- Be easy and safe to re-heat.
- Be original.
- Serve one person – an individually packaged packed lunch product.
- Be easy to eat away from home.

E *Design criteria*

Summary

You should now be able to:

carry out relevant research

carry out research analysis

use research analysis to write detailed design criteria.

AQA Examiner's comment

Jason has listed the results he found from carrying out each type of research. He has a wide range of results that he uses to write his design criteria. He could have included some of the more specific results from his research to narrow down the types of products he could choose as possible ideas to develop.

AC5 communication

- Good detailed explanations of information gathered with explanations presented in a clear and coherent manner.

- Text is legible, easily understood and shows a good grasp of grammar, punctuation and spelling.

AQA Examiner's comment

Jason has used the information he gathered through the research to write his design criteria. He has listed the general design criteria which will be easy to evaluate against. The criteria allow him scope to choose a range of initial ideas to investigate before he chooses one to develop.

AC5 communication

- Good detailed bulleted list – easy to understand.

- Concise and relevant information.

Development and investigations to produce a design solution

Development 1

After you have decided which product or prototype you are going to develop it is important to plan the types of development and investigational activities you could carry out to ensure you get an exciting new product. It is important to get your thinking written down on paper and the plan of developments can be in any format to help you focus your thoughts, e.g. spider diagram, bulleted list.

Developments need to consider the whole product. Sometimes these have different parts, e.g. lasagne; this type of product lends itself well to a range of different developments.

You need to consider the working characteristics of the product, i.e. how do the different ingredients work together to make that product. If you change the proportions of the ingredients or the types of ingredients in a product how is the product changed, e.g. do you get the same rise in a baked product if you use plain flour instead of self raising flour. What effect does this have on texture, taste, appearance?

Developments need to be more than modifications.

When presenting evidence of development in your folder you should include:

- what you are aiming to find out
- how you will carry out the investigation
- how you will use the results from the investigation to make decisions for your final product.

Chelsey's task (see Picture **A**) is to produce a cook/chill main course product and she is at the stage of developing a pasta product. She is on her second development where she is considering how to develop the colour and flavourings of fresh pasta using natural fresh ingredients. She has made one quantity of fresh pasta, divided it into four equal amounts and added her range of flavourings. She photographs various stages of the investigation to help provide evidence.

AQA Examiner's comment

Chelsey has most of the information needed for a development sheet. She has stated what she aims to find out and why. She lists the ingredients she is going to use and their functions in the correct context. She has annotated photographic evidence of the making she has carried out, and she has used two types of sensory analysis – even though these are of a simple nature they help inform the evaluation. She has a detailed evaluation that links her findings to her product specification; she could then have stated how she intended using these findings to carry out her next development.

AC5 communication

A good example of how to set out a page – all the information is clear and concise. Photographs are annotated and there is good use of ICT. The text is easily understood and there are high standards of grammar, punctuation and spelling.

Jenny is developing a fruit tartlet that is made up of four distinct components: the pastry case, the custard style filling, the fruit topping and the glaze. It is a product that has lots of scope for development. Each making session she does not make every part over and over again but considers a different part of the product, see source **B**.

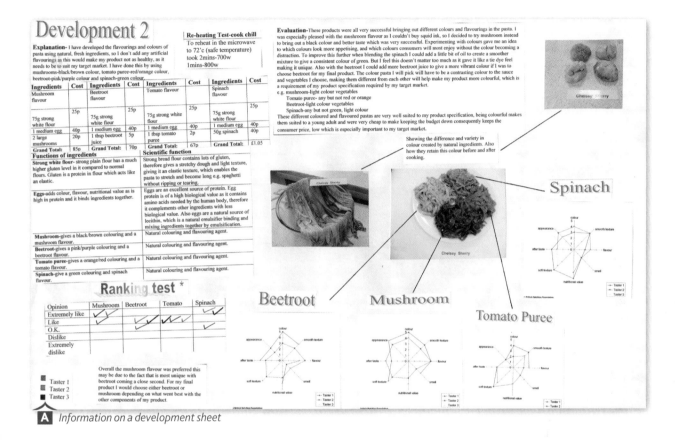

A *Information on a development sheet*

You must also:

- evaluate your results against your product specification
- consider if you can use your results to develop your product further.

You must try to:

- use small quantities
- include CAD if possible
- carry out fair testing.

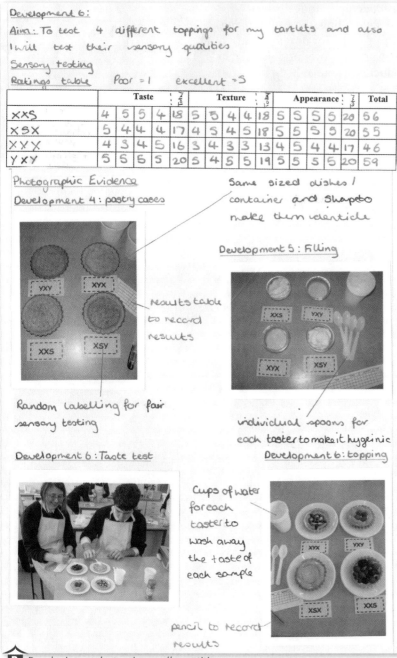

Development 6:

Aim: To test 4 different toppings for my tartlets and also I will test their sensory qualities

Sensory testing

Ratings table   Poor = 1   excellent = 5

	Taste				Tot	Texture				Tot	Appearance				Tot	Total
XXS	4	5	5	4	18	5	5	4	4	18	5	5	5	5	20	56
XSX	5	4	4	4	17	4	5	4	5	18	5	5	5	5	20	55
XXX	4	3	4	5	16	3	4	3	3	13	4	5	4	4	17	46
YXY	5	5	5	5	20	5	4	5	5	19	5	5	5	5	20	59

Photographic Evidence

Development 4: pastry cases

Same sized dishes / container and shape to make them identicle

Development 5: Filling

results table to record results

Random labelling for fair sensory testing

individual spoons for each taster to make it hygeinic

Development 6: Taste test

Development 6: topping

Cups of water for each taster to wash away the taste of each sample

pencil to record results

B Developing products using small quantities

Ruhena is developing a sweet pastry product and has carried out a series of developments each using small quantities, see Picture **C**. These show high level making skills at each stage of the development.

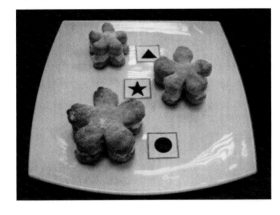

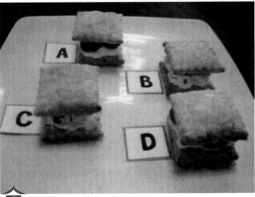

Ruhena has carried out investigations into six different aspects of her dessert product. She has considered the type of pastry, filling and topping, as well as the functions of ingredients, the sensory characteristics, the shape, the size and the finish. She has had an aim for each investigation, she has carried out sensory evaluations, considered the functions of the ingredients she is using and a conclusion. Her conclusion has helped her plan her next investigation.

AC5 communication

This is an excellent example of high quality presentation. The language is appropriate, easily understood and technical language is used correctly.

As part of Jane's developments into cold desserts she realised that she would need to consider various factors that could affect the setting properties of the desert. Jane decided to carry out an investigation into setting agents. Higher level students need to consider the working characteristics of the ingredients they use when developing products.

Sample	Easy to Use?	Consistency	Appearance/Comments
✡ Vege-Gel (Vegetarian Option)	Yes, however it is tricky to get it to work properly if it is not given 100% effort.	Awful; very lumpy and stodgy both during cooking and when turned out of its mould.	Uneven and lumpy; when the broken apart by a spoon, its interior was very mashed and broken.
Y Hartley's Powdered Gelatine Mix	Very easy; clearly stated method printed on the back of the packaging.	Strong and firm, however it did tend to get blemished very easily when it came out of its mould.	A very rich, dark purple colouring (beetroot red natural colouring). Difficult to see the raspberries inside. Smooth edges, few imperfections.
K Instant Gelatine CONTROL	It went rather lumpy and did not completely dissolve in the warm liquid. The remaining grains of gelatine were very difficult to remove from its equipment.	It formed a large crack through its centre which may have been due to the ratio of gelatine to liquid or that the product had been removed from its mould incorrectly.	A light translucent jelly was made, and the cranberry juice gave it a pleasant colour. The raspberries were easily visible. Ruined by the large split through the middle.
★ Leaf Gelatine	Extremely. It melted into the hot liquid with little mixing required.	Very pleasing; it was firm and turned out of it's mould easily. It had an idyllic structure; not too firm or too soft.	A lovely smooth texture and even colour. Firm and looks incapable of collapsing.

D *Functions of ingredients*

Tom, see Picture **E**, is developing a cheesecake product and after developing the base, cheesecake filling he is deciding upon how and what to use to decorate the top of the cheesecake.

Orange Segments
This cheesecake would have 4 orange segments facing the same way at different angles. This is quite simple and looks good.

Orange Segments in Chocolate
This cheesecake would have 4 orange segments ½ covered in melted chocolate facing the same way at different angles. This is a little less simple but still looks good.

Chocolate
This cheesecake would have melted milk chocolate covering the top of the cheesecake with grated white chocolate on top of this. This is quite different and I think it looks good.

Piped Cream
This cheesecake would have piped cream around the edge. This is quite simple and looks good although I think the middle may be a little empty.

Orange Coulis with Fruit
This cheesecake would have orange coulis spread over the top (as with the chocolate) and would have 4 orange segments facing the same direction at different angles. This is a little more difficult but looks good.

Grated Zest with an Orange Slice
This cheesecake would have orange zest grated on the top and a slice of orange arranged on top of this. This is more minimalist but looks good.

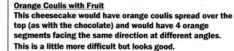

E *How to use Simple CAD in development*

Garry, see Picture **F**, has completed his initial development for his brief, which was to make a main meal product to be sold in the chilled section of a supermarket. He has developed a curry. He has researched types of accompaniments, made a range of ideas and evaluated each one. He has evaluated these against the product specification and has decided to develop a bread based product.

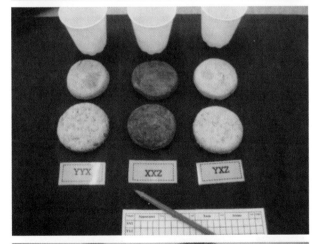

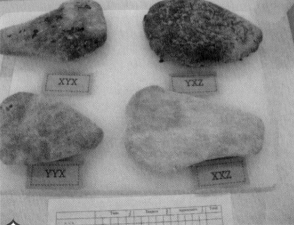

Garry has carried out three investigations on bread products. These developments were all carried out using small quantities and by changing some aspect of the bread based accompaniment. He has carried out changes to the flours, flavourings and added ingredients to alter the texture. The photographs show the products being sensory tested. Garry evaluated his results against the specification and drew conclusions.

Summary

You should now be able to:

develop an existing product

carry out investigations into ingredients and materials to produce a final product.

F *Possible developments of an accompaniment to curry.*

17.3 Making: high-level skills

The amount and quality of all making that is carried out accounts for up to 50% of your final design and making assessment marks. It is very important therefore to include a wide range of making skills at all stages of the design process. You must ensure they are carried out to a high standard.

All products should have a high quality finish, include a wide range of making skills, and be of a standard that a consumer would want to buy and eat. It is also encouraged that you include photographic evidence of making. This should include evidence of all the products you have made. It is also useful to include photographs of not just the finished product but also the stages of its making.

Caroline, see Picture **A**, is considering what products she could develop and has included a range of high level making skills through her design ideas. Her brief is to make a healthy option family dessert high in Vitamin C.

DESIGN IDEA 2: Orange and ginger pancakes

Ingredients; 125g plain flour, 2 eggs, beaten, 25 cl milk, 1 tbsp caster sugar, 30g butter + 30g for frying, 1 pinch of salt, 2 oranges, 1/2 pot of marmalade, powdered ginger, icing sugar

DESIGN IDEA 3: Lemon Meringue Pie

Ingredients; 180 gms (6 ozs) plain flour, sifted,pinch of salt,100 gms (3 ½ ozs) chilled butter, 2 tablespoons caster sugar, 1 egg yolk lightly beaten, 1 – 2 tablespoons iced water, 60 gms (2 ozs) corn flour, 150 gms (5 ozs) caster sugar
125 ml water, Finely grated zest of lemon, 125 ml lemon juice, 4 egg yolks, 30 gms (1 oz butter),
4 egg whites. ¼ tsp cream of tartar. 150 gms (5 ozs caster sugar)

DESIGN IDEA 4: Fromage Frais Cheesecake with Strawberry Sauce

Ingredients; 12 oz (350 g) fromage frais,12 oz (350 g) full-fat curd cheese, 3 large eggs, 6 oz (175 g) caster sugar, 1 teaspoon pure vanilla extract, 1½ lb (700 g) hulled strawberries, 2 level tablespoons caster sugar, 6 oz (175 g) sweet oat biscuits, 2 oz (50 g) melted buuter, 2 oz (50 g) coarsely chopped toasted hazelnuts.

A *High quality making during initial design ideas*

AQA Examiner's comment

Caroline has made three existing products and included some high level making skills such as short crust pastry, meringue, a blended sauce, a cheesecake and even-sized pancakes. It can be seen from the photographs that the presentation and finish of all her making is good. When recording her making she has included the methods for her products; this is not necessary.

David is making a product that could be included in 'the finest' range at his local supermarket. He is developing his own pasta.

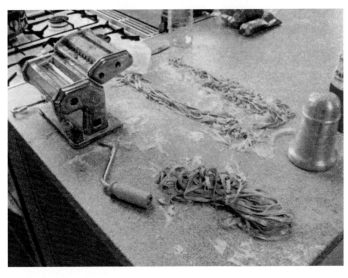

B   *The stages in making a product*

AQA
Examiner's comment

David has included photographic evidence of him carrying out the processes involved in making pasta. He shows us the finished product. Ideally his photographic evidence would have been annotated. The quality of his pasta looks good from the photographs.

AC5 communication

Good photographic evidence showing the stages of making.

Carla's brief was to develop a packed lunch product and after researching and carrying out other investigations she is now developing a pastry pasty product, see Picture **C**.

Plan: The aim for this experiment was to compare three different types of pastry to see which ones gave the correct colour, flavour, texture and had moisture to be used as the pastry for cheese and onion pasties. I also took note of the time it took each one of the different pastries to cook and I have used the results to come to my conclusion.

The pastry types: 1) Filo pastry
2) Flaky pastry
3) Shortcrust pastry

Pictures: The pictures at the top, to the right, show the different pastry types once cooked, but still whole. The pictures at the bottom show the pastry after being cut open once with a knife. These pictures demonstrate the 'mess' made by each pastry type.

Filo pastry   Flaky pastry   Shortcrust

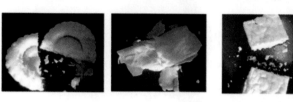

Pastry Type	Cooking Time	Colour	Description once Cut	Suitable for Pasties?
Filo	4 minutes 20	Yellow	Hard and crispy. Large, hard flakes created once cut. Dry.	No
Flaky	4 minutes	White/Yellow	Smooth texture and look. Small, light flakes created. Not too dry.	Yes
Shortcrust	4 minutes 40	Pale Yellow	Crumbly. Lumpy looking. Creates small flakes when cut but is very dry.	No

Filo	Person				
	A	B	C	D	E
Appearance	2	2	1	2	2
Colour	3	2	3	4	3
'Mess'					
Texture	4	4	5	3	4
Moisture	3	3	4	4	3

Flaky	Person				
	A	B	C	D	E
Appearance	1	1	1	2	1
Colour	2	1	2	2	2
'Mess'					
Texture	2	2	1	1	1
Moisture	2	2	2	2	2

Shortcrust	Person				
	A	B	C	D	E
Appearance	3	3	3	4	2
Colour	1	1	2	1	2
'Mess'					
Texture	2	2	3	1	3
Moisture	4	5	4	5	3

Product Specification Point Met: The pasties must have a light and flaky pastry that has moisture, creates some flakes when eaten but will not create mess when eaten.

Good examples of quality control in action whilst making are good to have in your coursework

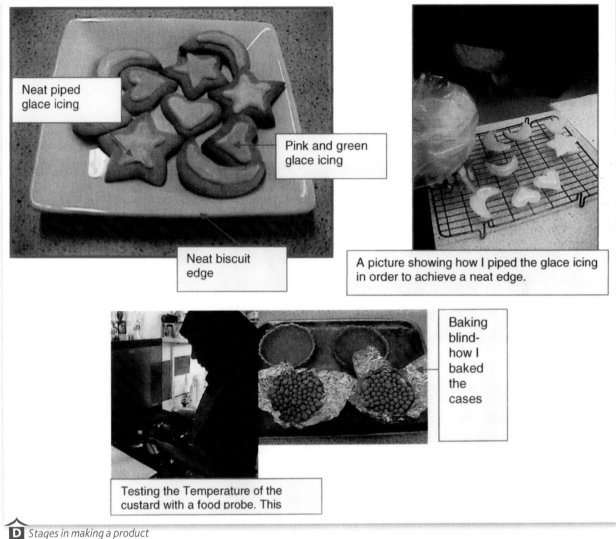

Neat piped glace icing

Pink and green glace icing

Neat biscuit edge

A picture showing how I piped the glace icing in order to achieve a neat edge.

Baking blind- how I baked the cases

Testing the Temperature of the custard with a food probe. This

D *Stages in making a product*

AQA
Examiner's comment   Mulheena (Picture D) is showing how she used a temperature probe to test that the custard had been heated to the required temperature. Also how to get a good texture with her pastry by baking it blind. She is decorating biscuits to a high standard.

AC5 communication

Good annotated photographic evidence.

A variety of students have produced the following products to show their high level making skills in the design ideas section of their controlled assessment. The brief was to make a vegetarian product.

 Examiner's comment These savoury products show high level making skills and finish. The photographic evidence ensures that an accurate record of the standard of making is recorded.

E *Vegetarian final products*

A variety of finished products showing accuracy in making, good quality control and high quality finish.

F *High-level making skills*

Summary

You should now be able to:

use a range of high level making skills

make products with high quality appearance and finish.

17.4 Production planning

A production plan is only required for your final product. It will replicate what would be needed to make your product in a test kitchen and not for mass production in a factory. You must include information to ensure that your product is made in a safe and hygienic way and that quality control checks are in place to ensure a high quality final product. You do not need to include information on risk assessment systems and control, e.g. HACCP. This information may be assessed in your written exam.

Eve's task was to develop a product suitable for a consumer who had coeliac disease. After developing her final product Eve has written a production plan to ensure a high quality product is made in the test kitchen, see Picture **A**.

Objectives

Understand how to write a production plan.

Understand the need for quality controls on a production plan.

AQA Examiner's comment

This is a detailed production plan laid out as a flow chart. Each making stage on the plan is stated and the quality checks that will be carried out at that stage also listed. Exact measurements, references to consistencies, and correct terms for each making process have also been stated. The quantities and ingredients for the final recipe will be listed on another design sheet.

AC5 communication

A good example of appropriate language, concise presentation of information and good spelling, punctuation and grammar.

PROCESS	QUALITY CHECKS
START	
Put on hat and apron, and wash hands and surfaces using anti-bacterial soap.	Ensure that they have been transported correctly at the right temperature and are within best before/use by date.
Take delivery of all ingredients	Check that all ingredients have been weighed correctly within their tolerance.
Weigh and prepare all the ingredients, following the manufacturing specification.	Check the biscuits have been crushed into fine breadcrumbs and that the butter is evenly mixed into the crushed biscuits.
Melt the butter in the microwave. Crush the biscuits and mix together.	Ensure that the mixture is 10mm thick
Place the biscuit mixture in the tin	Check that the ingredients are mixed thoroughly.
Mix together all the ingredients for the curd.	Make sure the mixture is thick enough to cover the back of a spoon, and that there are no lumps in the mixture.
Using a double saucepan, heat the mixture until its thick. for 5 mins	Ensure that the dried pineapple is stirred evenly throughout the curd
Remove from heat & chop and stir in the pieces of dried pineapple (5mm length).	Check there is 30mm of curd mixture.
Add the curd mixture on top of the biscuit base.	The curd should be stored at the correct temperature between 1 and 4°C
Refrigerate the curd	Ensure all the ingredients are mixed together well, and the mixture is smooth.
In a large mixing bowl, mix together all the ingredients for the cake mixture	Make sure that the cake mixture fills the rest of the tin, and that the mixture is smoothed out on top. to ensure an even topping
Pour the cake mixture on top of the curd	The cake should be golden brown, and the internal temperature above 72°C for 2 mins
Cook in the oven at a medium heat for 15mins	Check the cake is fully cooked, and the golden syrup is soft enough to pour.
Remove from the oven and melt the golden syrup	Check the golden syrup is spread evenly all over the cake topping, and there are 8 almonds (around 20mm in size)
Glaze the cake with the golden syrup and add the almonds	Check the product is packaged correctly, with no recycled packaging touching the product, and all the correct labelling information on it. Check that it is stored in the right storage conditions.
Package.	

A *Production planning*

Ruhena's task was to produce a sweet pastry dessert product. Her production plan is in the form of a flow chart and has photographic evidence alongside the plan, see Picture **B**.

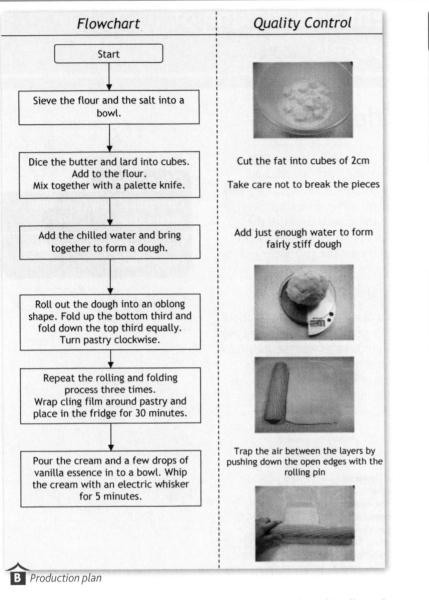

Flowchart	Quality Control
Start	
Sieve the flour and the salt into a bowl.	
Dice the butter and lard into cubes. Add to the flour. Mix together with a palette knife.	Cut the fat into cubes of 2cm Take care not to break the pieces
Add the chilled water and bring together to form a dough.	Add just enough water to form fairly stiff dough
Roll out the dough into an oblong shape. Fold up the bottom third and fold down the top third equally. Turn pastry clockwise.	
Repeat the rolling and folding process three times. Wrap cling film around pastry and place in the fridge for 30 minutes.	
Pour the cream and a few drops of vanilla essence in to a bowl. Whip the cream with an electric whisker for 5 minutes.	Trap the air between the layers by pushing down the open edges with the rolling pin

B Production plan

AQA Examiner's comment
A very detailed flow chart stating each stage of athe making along with checks to ensure a high quality product is produced every time. Excellent use of photographs to show what quality checks were carried out and how they were carried out.

AC5 communication
Excellent use of ICT and photographs. The text is easily understood, accurate use of technical terms and good spelling, punctuation and grammar.

AC5 communication
The presentation is very clear and concise. Appropriate use and understanding of technical language is included.

Julia has produced meatballs in an Italian sauce with tagliatelle as her final product. The production plan includes quality checks and making processes, see Picture **C**.

•**Fresh Pasta:** Mix flour, egg and oil together and mix until moist and has a firm consistency	•Procedure: Mix pasta ingredients for 2 minutes on a high speed. If smooth consistency is not achieved continuing mixing at 10 second intervals.
• **Meatballs:** Mix turkey mince and cranberry sauce together. Add breadcrumbs and egg to bind the ingredients together. Roll into even size and shaped balls ready to be cooked.	• Check all the breadcrumbs are the same size so that they do not appear obvious in the meatballs. •Visual check: cranberry sauce and breadcrumbs are evenly mixed throughout the minced turkey. Weigh each turkey ball to be same weight of 50g, allow tolerance on 2g.
• **Pasta:** Roll out pasta using pasta forming machine. •Shape pasta into tagliatelle	• Visual check: Complete pasta layers should be formed. Check thickness of pasta. Final sheet must be 3mm in thickness. Continue rolling if this is not achieved until correct thickness.
• **Tomato sauce:** Puree tomatoes, heat for 2 minutes. Add in pre-cooked bacon chunks, grated cheese and mixed herbs. Continue to stir on low heat for a further 2 mins.	•All bacon chunks are cut to the same size, abiding by tolerances. •All grated cheese is grated using the same sized grating panel.

C Production plan

AQA Examiner's comment
A detailed chart showing each stage of making described in lots of detail. Against each section of making are the quality checks that Julia carried out to ensure a high quality product. Reference to acceptable tolerances for size of ingredients and components such as the meatballs are stated.

Summary

You should now be able to:

write a production plan

apply quality control checks to a production plan.

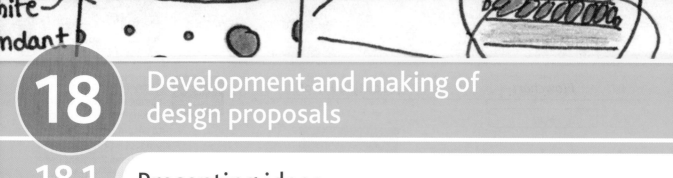

18.1 Presenting ideas

Design ideas are produced as a result of carrying out research and writing design criteria. The way they are presented can take many forms. Design ideas should provide a range of different examples of food products which fit the design criteria and take account of the research you have carried out. Design ideas should give information about the original product but also wherever possible give ideas of how a product could be developed.

The following are some of the ways design ideas can be presented.

Jack produced this idea as one of ten from a design task which asked for a savoury low cost main meal product. Jack's research had indicated that pasta was a nutritional, popular low cost ingredient as was the minced beef and tomatoes for the meatballs. In addition Jack had written in his design criteria that the product must contain pasta, a protein, some vegetables and have an interesting flavour. It should also cost no more than £1.50 and be suitable for a family meal, see Picture **A**.

> **Objectives**
>
> Understand how to present design ideas.
>
> Understand how to annotate design ideas.

> ∞ links
>
> See page 106.

Design idea 6: Spaghetti and Meatballs

Circular shaped beef meatballs with herbs and onions. The herbs and onions will vary the texture and make the meatballs more appetising by adding a higher quality flavour. Again, the meat will act as a high protein food.

Egg flavoured spaghetti as the base of the meal. It will be a standard component and will have a soft texture. This will be a neutral flavour that compliments and enhances the meat and spicy tomato sauce. The spaghetti provides a high carbohydrate element to the meal making it satisfying and long lasting.

A distinct spicy tomato sauce covering the meatballs, to make them moist and highly flavoured. There will also be chopped onions, garlic and oregano in the sauce to help give extra Italian flavouring.

A Presentation of Spaghetti and Meatballs

> **AC5 communication**
>
> This idea is presented well, it is clear and the text is accurate, concise and easy to read and understand.

AQA Examiner's comment
Jack has used a photograph of spaghetti and meatballs as the centrepiece for this idea. This gives a visual focus to the idea and is a good strategy if your sketching is not good. He has included a very descriptive explanation of the product which included both ingredient detail and nutritional information. The explanation implicitly includes the points he has made in design criteria. He does mention that the spaghetti is a standard component which is acceptable for this idea; however should he decide to develop this idea he needs to 'make' pasta – possibly different shapes and sizes, using different flouts, etc. To improve this design idea Jack could have made suggestions for developing the pasta, the sauce and the meatballs.

Ruhena has produced this idea as one of eight from a design task which asked for a product to be served at a party, see Picture **B**. She used her findings from research and analysis to write her design criteria which included the following points:

- a savoury or sweet product.
- contains fruit or vegetables.
- luxury range.
- attractive and appealing.

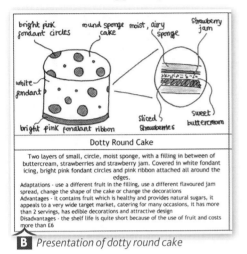

 Presentation of dotty round cake

Charlotte is working to a design task which asked her to design and make a cook-chill main course product, see Picture **C**. This is one of eight ideas she produced which took into account the information she gathered from her research and the points she had made in her design criteria. Her criteria stated that the product should be nutritious, contain protein, carbohydrate, a vegetable and could be frozen. The criteria also stated that the product should be moist and be tasty.

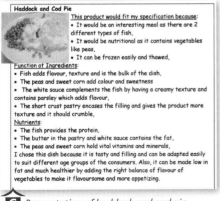

C *Presentation of haddock and cod pie*

AQA Examiner's comment

Charlotte has used pictures from the internet to illustrate her idea. These give a visual description of what she is considering. Her annotation not only gives an additional description of the idea but also includes a short evaluation against the design criteria. Another good feature of this idea is that she comments on the nutritional content of the product in some detail and also the functions of the ingredients. She will be able to use this information as a starting point for the development of the haddock and cod pie.

AQA Examiner's comment

Ruhena has chosen to sketch her design idea and her annotation gives additional information about the idea with some possible ideas for development which she calls adaptations. The points in her design criteria have been addressed. She could have improved the annotation by giving further ideas for developments, e.g. using different cake methods, using different fats, flours and sugars, adding flavourings to the cake, etc.

AC5 communication

The idea is clear and the annotation is legible, concise and relevant.

AC5 communication

This idea gives lots of information and it is presented clearly and concisely. The bullet points add to the clarity making it easy to read and understand. Underlining the different aspects also makes the presentation of the idea clear.

Summary

You should now be able to:

select appropriate presentation methods for design ideas

annotate ideas fully

identify possible development opportunities

present information about each idea clearly and concisely.

After you have evaluated all your suggested ideas you need to select which ideas to make. Making design ideas gives you lots of opportunity to show a range of different making skills. Therefore in addition to evaluating your ideas against the design specification you also need to think about which ideas will enable you to show good making skills and use a range of different recipes, methods and finishes. The ideas you make will provide a substantial amount of evidence for the 'making skills' part of your project. Ideas which you make enable you to choose and follow an existing recipe and produce a very good quality product. In order to provide evidence of what you have done and how your made product has been evaluated you need to present the information in your design folder. There is no set way of doing this, however you need to include the following:

- a list of ingredients used with quantities – but not the method
- a sensory evaluation of the finished product
- a list of the functions of the ingredients used
- a picture of the finished product
- possibly some nutritional analysis – if your design criteria included a particular nutritional point
- evaluation against the design criteria.

Objectives

Understand how to present information about the design ideas that have actually been made.

⚭ links

See page 108.

Idea 2 – Spinach and ricotta pancakes

Ingredients
1 cup flour
2 tablespoons baking powder
1 cup whole milk
1 tablespoon vegetable oil
1 tablespoon water
2 tablespoon butter
150g spinach, cooked, water squeezed out
150g ricotta
Pinch grated nutmeg

Method
1) Add all of the dry ingredients and mix well.
2) Then add spinach and ricotta.
3) Add butter in a hot pan and add bits of the batter to it so it forms circles.
4) Cook the pancakes until they are golden brown on both sides.

Ingredient	Sensory Function	Nutritional Function	Physical Function
Flour	Flavour Bulking ingredient	Carbohydrate	Structure
Baking powder	Lightness to baked products, introduces gases	Carbohydrate, Vitamins	Raising agent
Milk	Binds dry ingredients, gelatinisation of starch	Vitamins especially calcium	
Vegetable oil	Flavour, moistness	Fat	
Water	Moistness		
Butter	Flavour, moistness	Fat	Shortens baked mixtures
Spinach	Colour, Flavour, Appearance	Vitamins	Filling
Ricotta	Flavour, Moistness	Fat	Filling
Nutmeg	Flavour		Decoration, appearance

Pros
- The mixture is easy to make.
- The filling for the dish is easy to prepare and cook with the pancake filling.
- The dish is colourful and would appeal to the younger age group.
- The dish contains spinach which is high in iron and this is an important nutrient for vegetarians.
- The pancakes take very little time to cook.

Cons
- When cooking the pancakes it is hard to get a definite shape.
- The consistency of the mixture needs to be right for the dish to work.
- There is a high amount of butter in the dish which means that the fat is higher and this is unhealthy for people to eat on a regular basis.

Adaptations
To change this recipe I would:
- Make the dish smaller so that it can be easy to eat.
- Try to reduce the amount of butter and fat in the dish to make it nutritionally healthier such as adding less ricotta cheese as this contains a lot of fat. To help to replace the ricotta cheese a healthier cheese could be found which has less fat in it.

(Star diagram with axes: Taste, Appearance, Cost, Target group, Savoury, Nutritional value, Transport, Portion size; scale 0–10. Legend: Ideal profile, Product profile)

Pancake structure
Spinach and ricotta warm filling

I have rejected this idea as I do not think it is suitable for my target market

A *Spinach and ricotta pancakes*

AQA Examiner's comment

Simone had given us a very clear picture of what she has made both visually and descriptively. She has included a list of ingredients and a method (which is not required). However she gives a lot of detail on the functions of the ingredients and should she decide to develop this product she could use this information to inform developments. Her evaluation included advantages and disadvantages of this idea and together with a product profile (star diagram) shows the actual sensory attributes compared to the desired sensory attributes. She does mention some possible developments but these are quite simple and lack demand. The photograph gives very clear evidence of a high quality product in terms of its shape, colour, appearance and finish. The information she has provided gives her sufficient evidence to reject the idea in terms of her design criteria and the opportunities for development.

AC5 communication

The presentation of this idea is clear, concise and legible. It is easy to read, spelling and grammar are accurate. Space is used well.

Simone has made this idea (see Picture **A**) as one of six. The design brief she is working to asks for a high-energy lunchtime snack for office workers. Simone chose to make this idea because it showed different skills to the other products she was making. The other products were: asparagus and goat's cheese parcels, chickpea and bean tortilla wrap, wholemeal bread rolls with a spicy tuna filling, individual savoury flans and pitta bread pockets. All these ideas used different making skills and in most cases many higher level making skills.

Katie has selected to make chocolate brownies and ice cream as a dessert as part of her solution to a design brief asking for a cold dessert, see Picture **B**. Other ideas she made were: lemon meringue pie, cheesecake, apple strudel, coffee gateau and profiteroles. You will see from this list that Katie chose very skilful products to make for her ideas in order to try to achieve a high grade overall. Also all the products used quite different skills. She decided to add ice cream to the brownies because it added extra skills and also transformed what is traditionally regarded as a cake into a dessert.

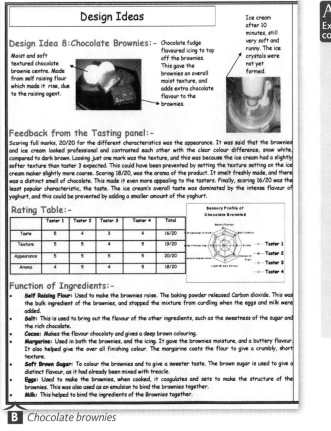

B *Chocolate brownies*

Katie has achieved conciseness by combining a list of ingredients with their functions to prevent repetition. Her sensory evaluation results enable her to give a detailed conclusion and by using two different methods of sensory evaluation, i.e. rating tests and product profile tests, she can combine the results to suggest improvements. It is clear that she has used several tasters and her conclusions are objective.

The photos show high quality making in terms of appearance and finish. This is supported by the comments from sensory testing indicating that the product was high quality in all aspects. By including the photo of the ice cream maker and the comments made about her ice cream, she gave clear evidence that she had made her own ice cream rather than using a ready-made product.

There is a lack of suggestions for developing this product which would have been very helpful in understanding if and why she rejected or selected the idea for development.

AC5 communication

This is an excellent example of appropriate language, concise presentation of information, accuracy, clarity and legibility.

Thomas has included this as one of his 'made' ideas to link to a design task asking for a savoury product which would reflect a particular culture, see Picture **C**. Other ideas Thomas made were: steak and kidney pie, lasagne, chicken korma and rice, pizza and enchiladas. Some of these ideas were more demanding than others. He used standard components in the enchiladas and in the korma which reduced the opportunity for 'making' skills.

Idea 5 – Chicken in Black Bean Sauce Stir Fry
Ingredients

4 chicken breasts, chopped
1 tbsp light soy sauce
1½ tbsp rice wine
½ tsp salt
1 tsp sugar
1 tsp sesame oil
2tsp cornflour
2 tbsp groundnut oil
5cm ginger, chopped
2 garlic cloves, chopped
1 red pepper, chopped
1 green pepper, chopped
2½ tbsp salted black beans ,coarsely chopped
150ml chicken stock

The chicken and black bean sauce recipe originates from ancient China. Due to the expensive black beans in the ingredients the product would be classed as a luxury meal. If I choose to develop this product I will look into different forms of the black bean sauce.

Name	Appearance	Aroma	Taste	Texture	Mouthfeel	Aftertaste	Sauce	Overall
Taster 1	4	3	3	3	4	3	3	3
Taster 2	3	3	3	3	4	3	4	3
Taster 3	4	4	4	4	4	2	3	4
Taster 4	3	4	3	3	5	3	3	3
Taster 5	3	4	3	3	4	2	2	3

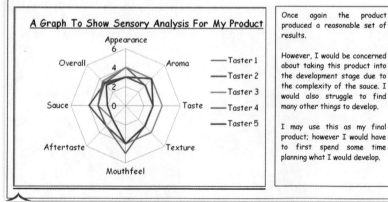

A Graph To Show Sensory Analysis For My Product

Once again the product produced a reasonable set of results.

However, I would be concerned about taking this product into the development stage due to the complexity of the sauce. I would also struggle to find many other things to develop.

I may use this as my final product; however I would have to first spend some time planning what I would develop.

C *Chicken in black bean sauce*

Ruhena has made this idea in response to her design criteria and her design task asking for a party product, see Picture **D**. Other products she made were: mini heart cakes, sausage and herb rolls, naan kebabs, valentine heart cake and garlic mushroom vol au vents. All these products are quite different and use a range of different skills and techniques which Ruhena carried out to a high level of accuracy and finish.

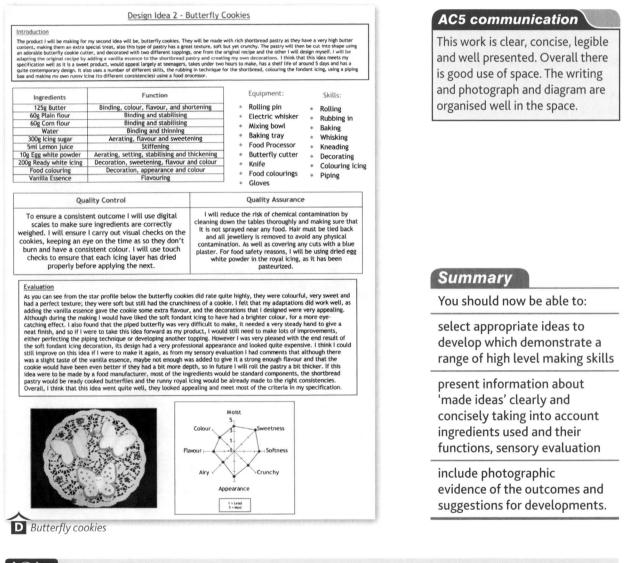

Design Idea 2 - Butterfly Cookies

Introduction

The product I will be making for my second idea will be, butterfly cookies. They will be made with rich shortbread pastry as they have a very high butter content, making them an extra special treat, also this type of pastry has a great texture, soft but yet crunchy. The pastry will then be cut into shape using an adorable butterfly cookie cutter, and decorated with two different toppings, one from the original recipe and the other I will design myself. I will be adapting the original recipe by adding a vanilla essence to the shortbread pastry and creating my own decorations. I think that this idea meets my specification well as it is a sweet product, would appeal largely at teenagers, takes under two hours to make, has a shelf life of around 5 days and has a quite contemporary design. It also uses a number of different skills, the rubbing in technique for the shortbread, colouring the fondant icing, using a piping bag and making my own runny icing (to different consistencies) using a food processor.

Ingredients	Function
125g Butter	Binding, colour, flavour, and shortening
60g Plain flour	Binding and stabilising
60g Corn flour	Binding and stabilising
Water	Binding and thinning
300g Icing sugar	Aerating, flavour and sweetening
5ml Lemon juice	Stiffening
10g Egg white powder	Aerating, setting, stabilising and thickening
200g Ready white icing	Decoration, sweetening, flavour and colour
Food colouring	Decoration, appearance and colour
Vanilla Essence	Flavouring

Equipment:
* Rolling pin
* Electric whisker
* Mixing bowl
* Baking tray
* Food Processor
* Butterfly cutter
* Knife
* Food colourings
* Gloves

Skills:
* Rolling
* Rubbing in
* Baking
* Whisking
* Kneading
* Decorating
* Colouring icing
* Piping

Quality Control	Quality Assurance
To ensure a consistent outcome I will use digital scales to make sure ingredients are correctly weighed. I will ensure I carry out visual checks on the cookies, keeping an eye on the time as so they don't burn and have a consistent colour. I will use touch checks to ensure that each icing layer has dried properly before applying the next.	I will reduce the risk of chemical contamination by cleaning down the tables thoroughly and making sure that it is not sprayed near any food. Hair must be tied back and all jewellery is removed to avoid any physical contamination. As well as covering any cuts with a blue plaster. For food safety reasons, I will be using dried egg white powder in the royal icing, as it has been pasteurized.

Evaluation

As you can see from the star profile below the butterfly cookies did rate quite highly, they were colourful, very sweet and had a perfect texture; they were soft but still had the crunchiness of a cookie. I felt that my adaptations did work well, as adding the vanilla essence gave the cookie some extra flavour, and the decorations that I designed were very appealing. Although during the making I would have liked the soft fondant icing to have had a brighter colour, for a more eye-catching effect. I also found that the piped butterfly was very difficult to make, it needed a very steady hand to give a neat finish, and so if I were to take this idea forward as my product, I would still need to make lots of improvements, either perfecting the piping technique or developing another topping. However I was very pleased with the end result of the soft fondant icing decoration, its design had a very professional appearance and looked quite expensive. I think I could still improve on this idea if I were to make it again, as from my sensory evaluation I had comments that although there was a slight taste of the vanilla essence, maybe not enough was added to give it a strong enough flavour and that the cookie would have been even better if they had a bit more depth, so in future I will roll the pastry a bit thicker. If this idea were to be made by a food manufacturer, most of the ingredients would be standard components, the shortbread pastry would be ready cooked butterflies and the runny royal icing would be already made to the right consistencies. Overall, I think that this idea went quite well, they looked appealing and meet most of the criteria in my specification.

D *Butterfly cookies*

AC5 communication

This work is clear, concise, legible and well presented. Overall there is good use of space. The writing and photograph and diagram are organised well in the space.

Summary

You should now be able to:

select appropriate ideas to develop which demonstrate a range of high level making skills

present information about 'made ideas' clearly and concisely taking into account ingredients used and their functions, sensory evaluation

include photographic evidence of the outcomes and suggestions for developments.

AQA Examiner's comment

Ruhena has presented clear evidence of her 'made' design ideas. She set out the context for her work and actually does begin to do some development work by changing the shape and the size of the cake from the original recipe. (It might have been better to make the original recipe at this point and develop it later as she will not be able to comment on the effect of the changes she has made to the original recipe.) In particular she can't comment on ratios and proportions of ingredients used, the need for reduction in oven temperature and time, etc. The information on functions of ingredients is very good and detailed, however there was no need to include a list of equipment used. It was pleasing to see quality control being included. Her comments were relevant and useful. She misunderstands Quality Assurance and really included health and safety which are not required at this stage.

Her evaluation is very thorough and she does comment on the sensory evaluation results – the star profile does not show how many people were involved which should have been included.

There is no real mention of the types of further developments she could carry out.

At this point she does not evaluate the ideas against the design criteria.

18.3 Product specifications

Product specifications are used to give details about the product that you have chosen to develop. A product specification relates to a specific product and will arise after you have evaluated design ideas. You will use the analysis of design ideas and previous information gathered from research to write your product specification. Although a product specification relates to one specific product it needs to be written in a way which will allow you to carry out lots of different developments

Samantha wrote this product specification (see Picture **A**) as a result of making a range of ideas for a product for a child's birthday party. Samantha evaluated her design ideas against the design criteria she had written. The idea she chose to develop needs to meet her product specification.

Objectives

Understand how to write a product specification which provides opportunity for a range of different developments.

⦾ links

See pages 116–117.

Ten Point Product Design Specification

1. Must be a sweet product which can be eaten sitting down at a birthday party
2. Target group of children 6 to 10
3. Red and green colours to be used as this is a finding from my research
4. Star shape as I found this out when trialling my ideas
5. Soft fillings as layered desserts such as cheesecake, millefeuilles and tartlets gained greater marks from sensory evaluations of my ideas
6. Each serving should be no more than 100g and 10cm in diameter as this is a good portion size for a child as I found from my research
7. Each serving must contain 1/5th RDA of vitamin C
8. Should be suitable for children with nut allergies and children allergic to artificial colours and flavours
9. Must include one or more of strawberries, kiwis, raspberries or peaches as these were the top fruits in my questionnaire
10. Price should not exceed 90p per serving

A *Product specification*

AQA Examiner's comment

This is a very detailed specification and uses information gathered from research and from the evaluation of design ideas. The specification gives scope for lots of development work. As Samantha works through the developments to her product she will be able to evaluate these against the specification.

AC5 communication

The presentation of the specification is very clear, concise and legible.

Having looked at my design specification, I have now decided on the specification points to develop and a few points have been removed from my original specification.

- It must be suitable for lacto-ovo vegetarians
- It must be able to be used as a lunch or dinner meal
- It must be healthy by containing at least three types of vegetable in the filling
- It must not be excessively high in fat and salt
- It must contain either a type of pulse or bean for high protein content
- The cost per portion should not exceed £1.50
- The weight should not exceed 400g
- It must be suitable for home freezing and microwave
- Before cooking, hygienic standards must be enforced: hair tied back, work surface wiped and clean equipment used where appropriate
- It should last at least two days in a chilled cabinet of a supermarket
- It should be suitable for batch production
- At least two ingredients in the finished product must have been British-grown (low food miles)
- Packaging must be recyclable, made of recycled products

B *Product specification*

AQA Examiner's comment

Sunita has produced a very good product specification showing good understanding. She makes very specific points. From what is written it is clear that information from analysis of research and ideas is well used.

AC5 communication

Very clear, concise, legible and relevant.

Sunita wrote her specification (see Picture **B**) in relation to a design task about vegetarian products. She had made a range of different ideas, some savoury, some sweet and evaluated these against her design criteria. This specification is very detailed using findings from her evaluation and also from her original research.

The next product specification was written by Mulbeena in relation to a design brief for cold desserts, see Picture **C**. A range of very different design ideas had been made which showed many different making skills. All the design ideas had been made to a high standard of appearance and finish. This product specification includes a photo of the finished product which is unusual as she added this after doing all her developments to how the product specification had been achieved.

Product Specification for a 'Fresh Cream Slice' type product

Aim: This specification gives me more specific features of my special occasion product idea. These must be retained or achieved during the development process.

- I will have a base layer made of pastry
- The pastry should have a light, airy texture and is golden brown in colour
- I must contain a fresh cream filling
- The product will contain at least one fruit
- It will be aimed at all age groups
- It will have a approximate cost of £4.50
- It will be sold in a pack size of 2, to cater for two individuals
- They will be sold and served chilled
- The products shelf life will be 3 days
- It should have good sensory qualities such as an exotic flavour, airy texture, mouth-watering aroma and sophisticated appearance
- It will be suitable to add other ingredients to sell as part of a range
- My product will have an attractive, bold label, with clear nutritional and dietary details
- My product will be mass produced

C Product specification

Lewis has written the product specification below from a design task relating to a main meal product for a family, see Figure **D**. He made three ideas but carried out very little evaluation. Therefore this product specification is not based on his findings from evaluation or research. It could actually be for many different types of products.

Product specification
- Is low in fat
- Is crispy and golden brown
- Is tasty and interesting
- Costs less than 50p
- Is soft in the middle
- Has a good aroma
- Attracts the attention of the customer
- Is produced hygienically and safely
- Is aesthetically pleasing
- Can be reheated easily with no loss of flavour

D Product specification

This is a very detailed specification and it is clear that it is based on a lot of knowledge and understanding. All the relevant points identified in analysis of ideas and research is included. The product specification is sufficiently flexible to allow a lot of development.

AC5 communication

Clear, concise, legible and relevant.

Lewis has produced a very general specification which lacks detail.

It is not clear how analysis of research and ideas has led to this product specification and it reads more like design criteria.

AC5 communication

Concise, legible but not totally relevant.

Summary

You should now be able to:

write a detailed product specification which uses information from analysis of research and ideas

enable a range of developments to be carried out

enable evaluation of each development to be made.

19.1 Evaluation of design ideas

The controlled assessment must 'tell the product development story' and therefore evaluation and testing must take place and be evidenced throughout your work. You need to thoroughly justify all the decisions you make. When evaluating it is important to take account of: the design criteria, results from sensory testing and the target group or intended consumer.

Two key areas should be evident when evaluating design ideas:

- After generating and presenting your design ideas, you then need to justify which products you will produce in the test kitchen; this is sometimes called 'concept screening'.
- After making four to six of your ideas you then need to decide and explain which idea will be taken forward to the development stage.

Selecting ideas to make in the test kitchen

Eve is designing and making a range of gluten free fruit desserts for consumers that suffer with coeliac disease. The products are to be sold in a major food retail outlet. From the research she discovered that there was a demand for both hot and cold desserts that included tropical fruits, particularly pineapple. Eve has generated a range of design ideas which were included on previous pages. She is using the design specification to decide and then justify which products to make in the test kitchen, see Picture A.

Justification of recipes chosen to be tested

I have chosen to make the eight products because they got the highest totals when I evaluated them against the design brief. The pineapple meringue pie is quite a fruity dessert that would not cost too much. It also covers some desirable criteria, such as looking attractive, tasting nice and it is quite low in fat. However, it is hard to make in an hour lesson, but can be managed. The pineapple upside down cake is very attractive and easy to make in the time available. It is a suitable size for a single portion and can be easily packaged in a paperboard box. However the cake part of the product is not very fruity, although the recipe can be adapted to make it fruitier. The roulade is a popular dessert and it was easy to adapt the recipe to make it gluten free. It would be safe to eat and it is simple to make it contain lots of pineapple, by using a pineapple curd rather than the more traditional lemon curd. The fritters are quite an unusual recipe, which makes them more interesting and appealing to buyers, as they are not already in the market. Also, they are quite cheap to make as they use tinned pineapple compared to fresh pineapple, which can be quite expensive. However, they are not very healthy as they are deep-fried but the criteria didn't specify that they had to definitely be healthy. Cookies are already popular in the market, but there are few gluten free cookies, and the pineapple flavour makes them more interesting. They are easily packaged and can keep for quite a long time. The pineapple ice cream with brandy snap baskets and pineapple sauce is a very fruity dessert. It uses a lot of skills to create a range of tastes and textures. There aren't many complete meal desserts already in the market so this would be suitable to be sold in a major food retail outlet. The pineapple-baked cheesecake is an interesting variation to the normal cheesecakes, as it is healthier (due to it being baked) and includes a layer of pineapple in the middle that is unusual. The mango pancakes are very fruity and contain mango as well as just pineapple. They are easy to make in a short amount of time although they do not cook for long, but could be sold in a major food retail outlet as a batter, or frozen so they can be cooked at home.

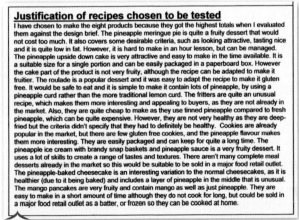

Design Proposals	The Product must be gluten free	The product must be suitable to be sold in a major food retail outlet	The product must be safe to eat	The product must be a dessert	The product must be fruity	The product must contain pineapple	The product must cost between £1 and £2	If labelled, the product must have all details on it needed by law	TOTAL
Pineapple Meringue Pie	5	5	5	5	4	5	4	5	38
Pineapple Upside Down Cake	5	5	5	5	4	5	5	5	39
Pineapple mousse	5	3	5	5	5	5	4	4	36
Saucy morello cherry puddings	5	3	5	5	5	3	2	5	33
Pineapple Roulade	5	4	5	5	4	5	4	5	37
Pineapple fritters with sauce	5	4	5	5	5	5	5	5	39
Pineapple baked Alaska	5	3	5	5	3	5	3	5	34
Pineapple cookies	5	5	5	5	4	5	5	5	39
Ice cream in brandy snaps	5	3	5	5	5	5	4	5	37
Pineapple baked cheesecake	5	4	5	5	4	5	4	5	37
Pineapple muffins	5	5	5	5	3	5	3	5	36
Tropical summer pudding	5	3	5	5	5	3	1	4	31
Mincemeat and pineapple tart	5	4	5	5	4	4	3	5	35
Mango pancakes	5	4	5	5	5	3	5	5	37
Pineapple rice pudding	5	3	5	5	4	5	4	4	36

Key
5 – good
1 – bad

Design Specification

A Evaluation of ideas

Selecting an idea for development

Alfie is designing and making a healthy option family dessert that has a high vitamin C content, see Picture **B**. He has made a very good range of design ideas in the test kitchen. All the ideas have been thoroughly evaluated on previous pages. He is now considering which product to take forward to the development stage. He has summarised the results using a table. Alfie's chosen idea to take to the development stage is pecan shortbreads with raspberries served with a raspberry purée.

Evaluation of Design Ideas

DESIGN IDEA	PLUS	MINUS	INTERESTING
Strawberry Gateau	High in vitamin C. Sponge base (as preferred in the research). It contains strawberries (also preferred in research) It scored 15/15 on the Hedonic rating test.	It takes a long time to make. Has a short shelf life.	The cake used could be developed by adding a flavour to it. The type of fruit could be developed to increase the vitamin C content. The decoration could be developed so it looks more appealing.
Orange and ginger pancakes	High in vitamin C. Contains oranges (as preferred in my research). Can be made on time. It serves 4 people. It has a sauce.	They might be difficult to vary the product. They are not a very interesting product. This product got one of the lowest marks on the Hedonic rating test.	The sauce could be developed to a flavour that people like more.
Lemon Meringue Pie	High in vitamin C. This product scored 14/15 on the Hedonic rating test. It looks attractive It has rich short crust pastry in.	It takes quite a long time to make therefore might not be practical to make it in the lesson.	The base could be developed, it could be given a flavour. The filling could be developed to create a more interesting taste. The topping could be developed to make it look more attractive.
Strawberry cheesecake	High in vitamin C. It contains strawberries. It looks attractive. It's suitable for families. It has a sauce.	It takes quite a long time to make.	The type of fruit could be developed to create a more contrasting taste. The base could be developed to create a more interesting texture. The filling could be developed to contain more vitamin C.
Shortbreads, Raspberries and Puree	High in vitamin C Contains Raspberries (as preferred in my research) It looks attractive. It's suitable for families. It has a sauce.	It takes quite a while to make.	The sauce could be developed to contain vitamin C. The biscuit could be developed to have a more interesting, sweeter taste. The fruit could be developed so that there is more vitamin C content. The filling could be developed. To have more of a contrasting taste to the fruit.
Orange and Lemon Refrigerator Cake	It is high in vitamin C. It contains more than one type of fruit, orange and lemon. It looks attractive. This product scored 14/15 in the Hedonic rating test.	It takes a long time to make, as it has to be left over night to freeze. Could be quite hard to develop it further.	The fruit could be changed to contain more vitamin C. The topping could be given a flavour to make the taste more interesting.
Fruit & Cream Puffs	It is high in vitamin C. It contains raspberries. It looks attractive. It is suitable for families.	This product got one of the lowest scores on the Hedonic rating test.	The pastry could be developed to make it more appealing to the consumer. The fruit could be developed to contain more vitamin C. A sauce could be included and developed to give a range of textures.
Rhubarb and Orange Meringue Pie	High in vitamin C Contains more than one type of fruit, orange and rhubarb. It looks attractive. It scored 14/15 on the Hedonic rating test	Not really suitable for families as rhubarb may be too sour for children.	The base could be developed, it could be given a flavour. The filling could be developed to create a more interesting taste. The topping could be developed to make it look more attractive.

Chosen Design Idea and Justification of Choice
My chosen idea - Pecan Shortbreads with Raspberries and Raspberry Puree
I have chosen this product because it is a dessert. It has a high content of vitamin C as it has a fruit that is high in the vitamin, so fits in with the design brief, more vitamin C could be added to increase the vitamin C content. It contains a fruit which could be changed to make it more popular to possible buyer, eg. put two types of fruit in instead of one as in my reseach I found out that people prefer tow fruits to one, although the raspberries that are currently in it were chosen by the people in my questionnaire as one of their favourite fruits. It is suitable for families as 8 are made, it is also hand held which makes it easier for children to eat. From my research I found that the consumers wanted a sauce with their dessert, my product has a sauce and this sauce could be developed further and changed. I could also develop the type of fruit used, the base type and the filling. This product looks attractive so would attract potential buyers to it, and it fits in the design specification. It scored as one of the highest in the Hedonic rating test, it was chosen as 15/15 so the people who tried it enjoyed it. The product does not contain large amounts of sugar, fats or salt, therefore making it quite healthy. It is suitable for all cultures except vegans as it contains egg and milk. I could use local products to reduce food miles making my product more environmentally friendly.

B *Selecting an idea for development*

AQA Examiner's comment Alfie has produced a concise table to communicate the positive and negative aspects of the ideas that were made in the test kitchen. Several of the points are repetitive and could have been further expanded. Alfie has considered possible developments for all the design ideas; this was included on previous pages so does not need repeating. At this stage it is only necessary to document the practical developments that will be carried out for the chosen product. Full justification of the chosen idea and possible areas for development are considered. Alfie has made good links to the research pages demonstrating an excellent understanding of the design process.

links
The criteria to evaluate each making activity can be seen on pages 158–159.

AC5 communication
There is excellent use of photographic evidence. The work is presented concisely. All the decisions are communicated in a clear and coherent manner using some technical language.

Summary
You should now be able to:

explain which ideas will be made in the test kitchen using the design criteria

fully justify which product will be taken to the development stage.

You need to provide the evidence of detailed testing and evaluation throughout the making stages. This can be achieved by setting up tasting panels and carrying out controlled testing. You should try to use a variety of sensory analysis techniques including: sensory profiling, rating and ranking tests. It is important that you constantly refer back to the specification and take account of the target group or intended consumer, when evaluating the success of a product.

Key occasions when sensory testing should take place are:

- after making your 4–6 design ideas
- when carrying out product appraisal
- after carrying out development activities
- when testing the final product.

Testing design ideas

Calvin is designing and making a range of soups with an accompaniment for a supermarket chain, see Picture **A**. He has chosen to test an Indian flavoured soup with naan bread. His target group is

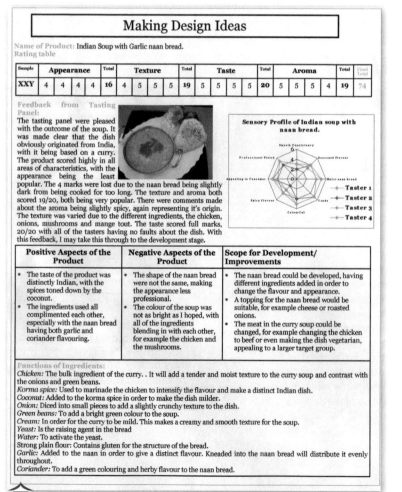

Making Design Ideas

Name of Product: Indian Soup with Garlic naan bread.
Rating table

Sample	Appearance				Total	Texture				Total	Taste				Total	Aroma				Total	Final Total
XXY	4	4	4	4	16	4	5	5	5	19	5	5	5	5	20	5	5	5	4	19	74

Feedback from Tasting Panel:

The tasting panel were pleased with the outcome of the soup. It was made clear that the dish obviously originated from India, with it being based on a curry. The product scored highly in all areas of characteristics, with the appearance being the least popular. The 4 marks were lost due to the naan bread being slightly dark from being cooked for too long. The texture and aroma both scored 19/20, both being very popular. There were comments made about the aroma being slightly spicy, again representing it's origin. The texture was varied due to the different ingredients, the chicken, onions, mushrooms and mange tout. The taste scored full marks, 20/20 with all of the tasters having no faults about the dish. With this feedback, I may take this through to the development stage.

Sensory Profile of Indian soup with naan bread.

→ Taster 1
→ Taster 2
→ Taster 3
→ Taster 4

Positive Aspects of the Product	Negative Aspects of the Product	Scope for Development/ Improvements
• The taste of the product was distinctly Indian, with the spices toned down by the coconut. • The ingredients used all complimented each other, especially with the naan bread having both garlic and coriander flavouring.	• The shape of the naan bread were not the same, making the appearance less professional. • The colour of the soup was not as bright as I hoped, with all of the ingredients blending in with each other, for example the chicken and the mushrooms.	• The naan bread could be developed, having different ingredients added in order to change the flavour and appearance. • A topping for the naan bread would be suitable, for example cheese or roasted onions. • The meat in the curry soup could be changed, for example changing the chicken to beef or even making the dish vegetarian, appealing to a larger target group.

Functions of Ingredients:

Chicken: The bulk ingredient of the curry. . It will add a tender and moist texture to the curry soup and contrast with the onions and green beans.
Korma spice: Used to marinade the chicken to intensify the flavour and make a distinct Indian dish.
Coconut: Added to the korma spice in order to make the dish milder.
Onion: Diced into small pieces to add a slightly crunchy texture to the dish.
Green beans: To add a bright green colour to the soup.
Cream: In order for the curry to be mild. This makes a creamy and smooth texture for the soup.
Yeast: Is the raising agent in the bread
Water: To activate the yeast.
Strong plain flour: Contains gluten for the structure of the bread.
Garlic: Added to the naan in order to give a distinct flavour. Kneaded into the naan bread will distribute it evenly throughout.
Coriander: To add a green colouring and herby flavour to the naan bread.

A Testing of a design idea – Indian soup with garlic naan bread

teenagers and he is therefore receiving feedback from this group. After making the design idea he set up a tasting panel to carry out sensory testing. To obtain reliable results the test was set up in a controlled way; arrangements included: serving the soup at the same temperature to all tasters, clear instruction for each tester, soups samples presented in identical sized and shaped containers, and clear response sheets to record the results.

Testing a development activity

Adele is designing and making a vegetarian product for the target group teenagers, see Picture **B**. She has taken lasagne through to the development stage. She is making and testing pasta because she was disappointed with the bland and unappetising flavour of the pasta used in the original dish. She has successfully made pasta using flour, egg and oil and is now testing the addition of flavourings.

⊂⊃links

Methods of sensory testing and recording techniques are explained in Chapters 13 and 15.

⊂⊃links

Further examples of evaluations can be seen on pages 158–159.

Product Development

Aim of the development
To develop the pasta with different flavours and colours to improve the sensory characteristics.

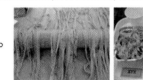

Tthe pasta dough is being extruded to make it possible to make the dough into tagliatelle.

Changes made with reasons
- When evaluating the original meal, there was an obvious lack of originality within the dish and therefore required developments in order to be improved.
- Through the making of three different flavoured pastas, I will be able to decide which pasta I will use
- The ingredients which I add in future developments will aid the originality of the product.
- The additional ingredients used helped compliment the taste of the rest of the product as pasta can be very bland. The extra ingredients added colour to the product, making the pasta more appealing.

Sample	Appearance				Total	Texture				Total	Taste				Total	Aroma				Total	**Total**
YXZ	4	3	3	3	13	5	5	4	5	19	5	4	5	4	18	5	4	4	5	18	68/80
YYX	4	3	3	4	14	5	4	5	5	19	3	2	2	3	10	3	3	3	4	13	56/80
XYX	4	4	4	5	17	4	4	3	5	16	4	4	4	5	17	5	4	4	4	17	67/80

How the changes affected the product
- **YXZ (Garlic)** - This was the most popular pasta type, scoring highly in all characteristics but the appearance. I feel this was due to the fact that the appearance was not changed by the garlic and looked like your average plain one. It had a distinct Italian flavour and strong aroma making it more appealing to the tasters. This will be the pasta I take forward and use in my final product.
- **YYX (Tomato)** - This pasta came out with a disappointing result but did score highly with the texture and was said to be very soft and moist. It was a deep orange colour due to the bright red colour of the tomato puree.
- **XYX (Mixed herb)** - This pasta was reasonably popular with the tasters. It scored mainly high on all aspects of it.

 Testing a development activity

AQA Examiner's comment

A clear aim has been established providing focus to the development activity. There is photographic evidence showing controlled testing procedures including: coded samples with random letters and water provided to cleanse the palate. Good photograph of the equipment used to ensure a consistent product is produced for each sample. Adele has explained why she has made the changes to the original pasta and the focus of the investigation. Very good use of a rating table to record the results of the sensory analysis but results (scores) must be referred to within the written text. She has explained how the changes affected the product and it is clear to see which flavourings have been tested. There is no conclusion to the activity; it is unclear which type of pasta will be used when the final product is produced. At the end of each activity it is good practice to explain the result and the next stage of the development process.

Objectives

Understand how to evaluate all aspects of the final outcome.

Understand how to produce and present nutritional information related to the final product.

∞ links

How to present nutritional information is shown in 15.2.

When all the development work has been carried out and evaluated it is time to make the final product. You are nearly at the end of the controlled assessment. When the product has been made and all aspects of the final outcome have been tested it is necessary to present a solution to the original design task.

The final design solution page should include the following:

- a photograph of the final product
- a full list of ingredients for all component parts
- a review of the development process including full justification of the choices made
- a comparison against the specification
- a brief explanation of how the outcome may need to be modified for commercial production
- final sensory testing results including comments made by the target group or client
- nutritional analysis of the product.

Mason has completed all the development work and evaluated and tested the final product (See Picture **A**). He is now in a position to present the final design solution. He has thought carefully about all aspects of the product and produced a very detailed final solution. Mason's brief was to design and make a pasta product for senior citizens.

AC5 communication

This is an excellent example of the use of ICT, appropriate technical language, concise presentation, accuracy and clarity.

AQA Examiner's comment

Mason has produced a final design solution. Comparison to the specification can be seen and this is presented concisely and all criteria discussed. He has fully justified why one of the specification points was not met. The page contains all the information required for the final design solution and is professionally presented.

A good section related to sensory analysis with the target group's opinion of the final product. Mason has selected some interesting criteria to evaluate the product against including: finish, ingredients used and quality control. It is not necessary to cost the dish unless this features as one of the design criteria. Social, environmental and moral issues discussed which is assessed as part of criteria 2. Limited reference to how the product would need to be modified for commercial production. Disappointing photograph of the final product – it is important that this shows the final outcome at its very best.

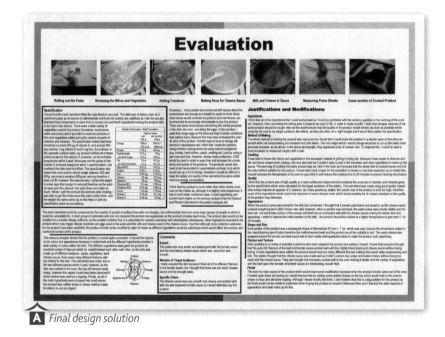

A Final design solution

Nutritional label

Jessica's design brief was to design and make a vegetarian food product suitable for a family that contains less than 15g of fat per portion. Jessica has decided to present the nutritional label on a mock up of the packaging, see Picture **B**.

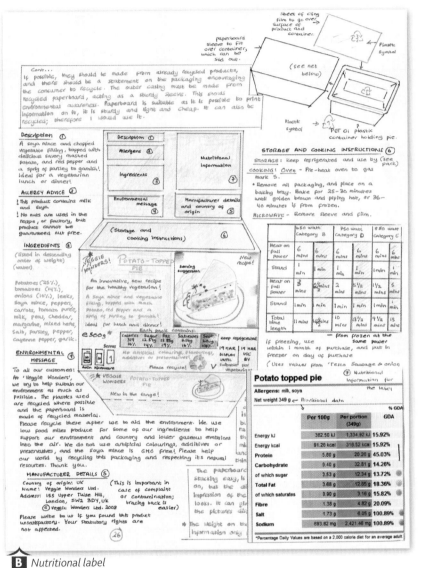

B *Nutritional label*

AQA Examiner's comment

Jessica has produced a net of the packaging to show how the nutritional label will be displayed. However, only a nutritional label is required. It is not a requirement to produce the packaging. The information is clear and nutritional analysis highlighted and this meets the requirements of the design brief. A nutritional analysis programme has been used to accurately calculate the amounts of each nutrient in the product. This was fully analysed on the previous page – final design solution. There is understanding of the information that should be included on a food label; this will help Jessica when answering questions on the examination paper. Good coverage of social, moral and environmental issues, this is assessed as part of criteria 2.

AC5 communication

The presentation is very clear and easily understood. Pleasing to see high standards of communication when ICT is not used. Text is easily understood and shows a good grasp of grammar, punctuation and spelling.

Summary

You should now be able to:

present a final design solution that contains all elements of the final product

analyse and present nutritional information.

Glossary

A

Acetic acid: vinegar.

Aeration: when air is trapped in a mixture.

Aesthetic: attractive.

Amino acids: the small units that form the chains in protein.

Anaerobic: not needing oxygen.

Analysis of brief: splitting up the design brief to identify key points.

Annotation: add explanatory notes.

Antibacterial: bacteria-hating substance that will usually kill bacteria.

Anti-oxidant: a substance that stops fat in food going rancid.

Appliance: a piece of electrical equipment.

Ascorbic acid: vitamin C.

Aseptic packaging: preserves foods without using preservatives or chilling.

Assembling: putting component parts together.

Attribute: particular characteristic of a food.

B

Biodegradable: broken down totally by bacteria.

Bland: lack of flavour/taste.

C

Caramelisation: process of changing the colour of sugar from white to brown when heated.

Citric acid: lemon juice.

Coagulation: when eggs are heated they change colour and become firm-set.

Colloidal structure: when two substances are mixed together.

Colloids: formed when one substance is dispersed through another.

Commercially viable: make a profit when it is sold.

Communication: pass on information, ideas and thoughts.

Consistency: ensures products are the same.

Consistent: the same quality each time a product is made.

Contaminate: to spoil or dirty something.

Couli: fruit that has been puréed, sieved and then thickened.

Critical Control Point (CCP): when a food safety hazard can be prevented/reduced to an acceptable level.

Cross contamination: the transfer of food spoilage/poisoning from one food to another.

Cryogenic freezing: food is immersed or sprayed with liquid nitrogen.

Curdling: fat separates from the sugar and eggs when the egg is added.

D

Descending: from the largest to the smallest.

Descriptor: a word describing a sensory characteristic, e.g. spicy.

Design brief: a statement which provides the situation for your designing and making.

Design criteria: a list of general points from which a range of different ideas can be made.

Deteriorate: starting to decay and losing freshness.

Development: make changes to a food product which will affect its characteristics.

Dextrinisation: when starch converts into a sugar.

Dietary Reference Values (DRVs): scientifically calculated estimates of the amounts of nutrients needed by different groups of people.

E

E numbers: the classification system used for food additives.

Eatwell plate: a healthy eating model, to encourage people to eat the correct proportions of food to achieve a balanced diet.

Emulsifier: a substance that stops oil and water from separating .

Emulsifying agent: a substance that will allow two immiscible liquids (substances that do not mix) to be held together, e.g. lecithin in egg yolk.

Emulsion: a mixture of two liquids.

Enrobing: coating and surrounding a product with another ingredient.

enzymic browning: reaction between a food product and oxygen resulting in a brown colour, e.g. sliced potato has brown patches when sliced and left in the air.

Estimated Average Requirement (EAR): the average need for a nutrient.

Evaluation: summarise information and make conclusions, judgements and decisions.

F

Fair testing: to compare like with like using only one variable.

Fermentation: when yeast produces carbon dioxide.

Finishing: completing the presentation of a food product to a high standard.

Foams: a mixture of gas and liquid is called foam.

Food additive: a substance added to a food product to improve its quality.

G

Gel: a small amount of a solid mixed in a large amount of liquid that then sets.

Gelatinisation: heated starch granules absorb liquid and swell, and burst to thicken liquid.

Genes: the basic unit of heredity used to pass on characteristics.

Gluten: protein found in flour.

Guideline Daily Amounts (GDAs): guide to the amounts of calories, sugar, fat, saturated fat and salt you should try not to exceed to have a healthy balanced diet.

H

Hermetically: airtight.

Higher level making skills: food preparation and cooking skills which require care, precision skills and understanding and which can be carried out to a high standard.

High-risk food: food which is an ideal medium for the growth of bacteria or microorganisms.

Hygienically: to prepare food in a clean to stop food spoilage or poisoning occurring.

I

Impermeable: cannot penetrate.

L

Landfill sites: large holes in the ground where bags of household waste are buried.

Listeria monocytogenes: a common food-poisoning bacteria.

M

Making skills: practical skills which show your ability to make food products.

Microorganism: tiny living thing that can only be seen through a microscope.

Mislead: not telling the truth.

Mnemonic: something to help us remember things.

Modification: simple changes which have little effect on the structure and composition.

Modified Atmospheric Packaging (MAP): used to extend the shelf life of food.

Modified starches: starches that have been altered to perform additional functions.

Monitoring: keep constant watch.

N

Net weight: not including packaging.

Non starch polysaccharide: the part of food that is not digested by the body.

Nutrient: the part of a food that performs a particular function in the body.

Nutritional analysis: using resources to find out the nutritional content of a product.

Nutritional content: the type and quantity of nutrients which the product supplies.

P

Pathogenic: causing disease.

Polysaccharide: dietary fibre.

Preservative: a substance that extends the shelf life of a food.

Preserve: to keep food fit to eat.

Prior knowledge: knowledge you already have without the need to find out.

Product analysis: examining a food product to find out the ingredients, packaging characteristics and properties.

Product specification: a list of features/characteristics/ properties which a food product must meet.

Profiling test: sensory evaluation test to identify individual specific characteristics of product.

Proportion: relative quantities of ingredients in a recipe, expressed in numbers.

Prototype: the first version of a product that is being developed.

Q

Quality control: steps taken to check a product at various stages of making to ensure a consistent and high quality outcome is achieved.

Questionnaire: questions asked to a range of people. Results can be used to inform ideas.

R

Raising agent: increases the volume of doughs, batters and mixtures by promoting gas release (aeration).

Recycled: to make into something else.

Reference Nutrient Intake: (RNI): the amount of a nutrient that is enough for most people in a group.

S

Sample: small amount of the product.

Scaling up: multiplying up proportionally.

Sensory analysis: identifying the sensory characteristics of products, i.e. taste, texture, appearance, mouth-feel, colour.

Sensory evaluation: using the range of senses to asses a food product – appearance, smell, taste.

Sensory qualities: the look, smell, taste, feel and sound of food products.

Shelf life: the length of time a food product can be kept and be safe to eat.

Shortening: when fat coats the flour particles preventing absorption of water resulting in a crumbly texture.

Solution: when a solid dissolves in a liquid, e.g. salt in water.

Specification: details which describe the desired characteristics of a product.

Standard component: pre-prepared ingredient that is used in the production of another product.

Standardise: make everything the same.

Staple food: a food that forms the basis of a traditional diet – wheat, barley, rye, maize or rice, or starchy root vegetables such as potatoes.

Suspension: a solid held in a liquid.

Sustainability: to continue to support.

Symptom: a sign of something.

Syneresis: loss of a liquid from a gel on standing, e.g. custard.

Tampering: to interfere with.

Target group: the specific group of people at which you are aiming your product.

Test kitchen: the place where a food technologist experiments and develops new products.

Tolerance: the amount of difference allowed when making.

V

Vacuum packaging: a way of preserving food that has been used for many years.

Vibrate: to move up and down very quickly.

Viscosity: the thickness of a mixture, e.g. sauce.

Visual: looking at something

Index

Key terms and their page numbers are listed in **bold**.